T0331252

# Plant Nutrition
*and*
# Soil Fertility Manual

SECOND EDITION

# Plant Nutrition

*and*

# Soil Fertility Manual

## SECOND EDITION

J. Benton Jones, Jr.

**CRC Press**
Taylor & Francis Group
Boca Raton   London   New York

CRC Press is an imprint of the
Taylor & Francis Group, an **informa** business

CRC Press
Taylor & Francis Group
6000 Broken Sound Parkway NW, Suite 300
Boca Raton, FL 33487-2742

© 2012 by Taylor & Francis Group, LLC
CRC Press is an imprint of Taylor & Francis Group, an Informa business

No claim to original U.S. Government works

Printed in the United States of America on acid-free paper
Version Date: 20111220

International Standard Book Number: 978-1-4398-1609-7 (Paperback)

### Library of Congress Cataloging-in-Publication Data

Jones, J. Benton, 1930-
   Plant nutrition and soil fertility manual / J. Benton Jones, Jr. -- 2nd ed.
     p. cm.
   Includes bibliographical references and index.
   ISBN 978-1-4398-1609-7 (pbk. : alk. paper)
   1. Plants--Nutrition--Handbooks, manuals, etc. 2. Soil fertility--Handbooks, manuals, etc. I. Title.

QK867.J66 2012
631.4'22--dc23                                                    2011050480

**Visit the Taylor & Francis Web site at
http://www.taylorandfrancis.com**

**and the CRC Press Web site at
http://www.crcpress.com**

# Contents

## SECTION I   Introduction and Basic Principles

# SECTION II   Physical and Physiochemical Characteristics of Soil

# SECTION III  Plant Elemental Requirements and Associated Elements

# SECTION IV   Methods of Soil Fertility and Plant Nutrition Assessment

## SECTION V    Amendments for Soil Fertility Maintenance

# SECTION VI    Methods of Soilless Plant Production

# *APPENDICES*

# Preface

Soil fertility and plant nutrition principles are the two primary subjects discussed in this book, presenting the reader with what would have been learned in basic as well as advanced soil fertility and plant nutrition college courses. The topics discussed are presented in such a manner that the reader can, with minimum basic background knowledge, feel confident applying the principles presented to his own soil/crop production system. The information in this book can be used as a means for searching particular topics by subject matter. In addition, this book contains sufficient fundamental information so that there is no need to search other sources unless there are specific issues associated with a particular soil-plant system that is not covered in this book, or more detailed factual information is desired.

The book is divided into two sections:

- Chapters that discuss the fundamental principles of soil and crop fertility management.
- Subject matter sections that deal with specific topical subjects.

## HISTORICAL BACKGROUND

Today, not many agricultural land-grant universities and colleges offer basic or advanced instruction on soil fertility and plant nutrition subjects applicable to the practical management of soils and crops. Research conducted at experiment stations and research centers within this land-grant system have in the past provided information needed by farmers and growers to keep pace with changing technology. This information was transferred through the Land-Grant Cooperative Extension Service in either written publications or by presentations given in seminars and instructional programs conducted by state specialists or local county agents. Although a portion of this system is still intact in some states, many land-grant institutions have significantly reduced their outreach programs dealing with soil fertility and plant nutrition subjects. Therefore, those who had relied on these institutions in the past have had to seek other sources for information essential for success in soil fertility management and crop production.

As soil and crop management procedures have become more complex, county agricultural agents, farm advisors, fertilizer and chemical dealers, as well as consultants have had to specialize in some aspect of soil fertility and crop nutrition management procedures, limiting their ability to provide a range of advice and services. Most farmers and growers can no longer turn to just one source for the information and instruction needed to achieve their production goals.

## DEFINITIONS AND JARGON

As with any subject, particularly one associated with a specific scientific field, there develops a jargon that is easily recognized by the practitioners but may be confusing to the learner and even those working in this field. For example, in the early 1950s, liquid forms of nitrogen-containing fertilizers were coming into use with "liquid nitrogen" as the defining words for such fertilizers. While listening to a farm advisory program that talked about how best to apply a "liquid nitrogen" fertilizer to one's lawn, a scientifically trained listener wondered how in the world one would apply liquid nitrogen, which to him, is a liquefied gas that has a temperature of minus 209°C, and would be both difficult and dangerous to handle without the proper equipment.

In this book, when there can be two or more possible meanings for a word or phrase, there is provided an explanation so that no such confusion occurs as was noted above with the use of the phrase "liquid nitrogen." Some of the common terms used in the past, and even in the current literature, can be confusing. For example, words such as nutrition, nutrient, nutrient element, essential element, mineral or mineral element, metal, etc., to some may be synonymous, or have a variety of meanings depending on the context. When such words are used in the text, their specific definition as applied to soil fertility and plant nutrition subjects is given. A glossary of general terms, together with those terms specific to the subject of soil fertility and plant nutrition, appears in Appendix A.

## ABBREVIATIONS

To make the text easier to read, appropriate and commonly used abbreviations are used unless there is a potential for confusion or misunderstanding in some contexts. Units of measurement, weight, volume, etc., are given mostly in British units or where the metric or other units are specific for that parameter, or that which is in common use. In most of the sections, tabular data are used with a minimum of verbiage, making the essential information more easily identified and applied. The following are the abbreviations used in the text for elements, compounds, ionic forms, and units of measure.

### Elements and Their Symbols

| Element | Symbol |
| --- | --- |
| Aluminum | Al |
| Arsenic | As |
| Boron | B |
| Cadmium | Cd |
| Calcium | Ca |
| Carbon | C |
| Chromium | Cr |
| Chlorine | Cl |
| Cobalt | Co |
| Copper | Cu |
| Hydrogen | H |
| Iron | Fe |
| Lead | Pb |
| Lithium | Li |
| Magnesium | Mg |

| Manganese | Mn |
|---|---|
| Mercury | Hg |
| Molybdenum | Mo |
| Nickel | Ni |
| Nitrogen | N |
| Oxygen | O |
| Phosphorus | P |
| Potassium | K |
| Selenium | Se |
| Silicon | Si |
| Sodium | Na |
| Strontium | Sr |
| Sulfur | S |
| Titanium | Ti |
| Vanadium | V |
| Zinc | Zn |

| Compound | Elemental Formula |
|---|---|
| Ammonia | $NH_3$ |
| Ammonium molybdate | $(NH_4)_6Mo_7O_{24} \cdot 4H_2O$ |
| Ammonium nitrate | $NH_4NO_3$ |
| Ammonium sulfate | $(NH_4)_2SO_4$ |
| Borax | $Na_2B_4O_7 \cdot 10H_2O$ |
| Boric acid | $H_3BO_3$ |
| Calcium carbonate | $CaCO_3$ |
| Calcium chloride | $CaCl_2 \cdot 4H_2O$ |
| Calcium nitrate | $Ca(NO_3)_2 \cdot 4H_2O$ |
| Calcium sulfate | $CaSO_4 \cdot 2H_2O$ |
| Carbon dioxide | $CO_2$ |
| Copper sulfate | $CuSO_4 \cdot 5H_2O$ |
| Diammonium phosphate | $(NH_4)_2HPO_4$ |
| Hydrochloric acid | $HCl$ |
| Ferric sulfate | $Fe_2(SO_4)_3 \cdot 4H_2O$ |
| Ferrous ammonium sulfate | $(NH_4)_2SO_4 \times FeSO_4 \cdot 6H_2O$ |
| Ferrous sulfate | $FeSO_4 \cdot 7H_2O$ |
| Magnesium carbonate | $MgCO_3$ |
| Magnesium sulfate | $MgSO_4 \cdot 7H_2O$ |
| Manganese oxide | $MnO$ |
| Manganese sulfate | $MnSO_4 \cdot 4H_2O$ |
| Monoammonium phosphate | $NH_4H_2PO_4$ |
| Nitric acid | $HNO_3$ |
| Phosphoric acid | $H_3PO_4$ |
| Potassium chloride | $KCl$ |
| Potassium nitrate | $KNO_3$ |
| Potassium sulfate | $K_2SO_4$ |
| Silica | $SiO_2$ |
| Sodium molybdate | $Na_2Mo_7O_{24} \cdot 7H_2O$ |

| | |
|---|---|
| Sodium nitrate | $NaNO_3$ |
| Sulfuric acid | $H_2SO_4$ |
| Urea | $CO(NH_2)_2$ |
| Zinc sulfate | $ZnSO_4 \cdot 7H_2O$ |

## Ionic Forms

| Element | Elemental Formula$^{Valance}$ |
|---|---|
| Aluminum | $Al^{3+}$ |
| Ammonium | $NH_4^+$ |
| Borate | $BO_3^{3-}$ |
| Chloride | $Cl^-$ |
| Calcium | $Ca^{2+}$ |
| Copper | $Cu^{2+}$ |
| Iron (ferrous, ferric) | $Fe^{2+}$ and $Fe^{3+}$ |
| Magnesium | $Mg^{2+}$ |
| Manganese | $Mn^{2+}$ |
| Molybdate | $MoO^{3-}$ |
| Nickel | $Ni^{2+}$ |
| Dihydrogen phosphate | $H_2PO_4^-$ |
| Monohydrogen phosphate | $HPO_4^{2-}$ |
| Orthophosphate | $PO_4^{3-}$ |
| Potassium | $K^+$ |
| Nitrate | $NO_3^-$ |
| Nitrite | $NO_2^-$ |
| Sodium | $Na^+$ |
| Silicate | $SiO_4^-$ |
| Sulfate | $SO_4^{2-}$ |
| Vanadate | $VO_4^{3-}$ |
| Zinc | $Zn^{2+}$ |

## Units of Measure

| Unit | Abbreviation |
|---|---|
| **Area** | |
| Acre | A |
| Hectare | h |
| Square meter | $m^2$ |
| **Volume** | |
| Cubic centimeter | cc |
| Liter | L |
| Milliliter | mL |

| Distance | |
| --- | --- |
| Feet | ft |
| Yard | y |
| Meter | M |
| Decimeter | dm |
| Centimeter | cm |
| Millimeter | mm |

| Weight | |
| --- | --- |
| Milligram | mg |
| Gram | g |
| Kilogram | kg |
| Pound | lb |

| Concentration | |
| --- | --- |
| Parts per million | ppm |
| Milligrams per liter | mg/L |

## THE INTERNET

Today the Internet offers a wide range of information, usually specific to a particular region, soil, and crop. Those turning to the Internet must be able to comb through numerous websites to find the information and instructions applicable to their particular soil/crop production system. In addition, they must be able to identity that information found to have application to their individual circumstances and conditions. One good clue for assessing the value of the information provided is to observe the date posted on the Internet and when last updated. The information in this book can be used for verification of information that has been posted on websites that either complements or adds additional information to the subject.

# About the Author

**J. Benton Jones, Jr.**, The author has written extensively on the topics of soil fertility and plant nutrition during his professional career. After obtaining a BS degree in agricultural science from the University of Illinois, he served on active duty in the U.S. Navy for two years. After being discharged from active duty, he entered graduate school, obtaining MS and PhD degrees in agronomy from the Pennsylvania State University. For ten years, Dr. Jones held the position of research professor at the Ohio Agricultural Research and Development Center (OARDC) in Wooster. During this time, his research activities focused on the relationship between soil fertility and plant nutrition. In 1967, he established the Ohio Plant Analysis Laboratory.

Joining the University of Georgia faculty in 1968, Dr. Jones designed and had built the Soil and Plant Analysis Service Laboratory building for the Georgia Cooperative Extension Service, serving as its director for four years. From 1972 until his retirement in 1989, Dr. Jones held various research and administrative positions at the University of Georgia. Following retirement, he and a colleague established Micro-Macro Laboratory in Athens, Georgia, a laboratory providing analytical services for the assay of soils and plant tissues as well as water, fertilizers, and other similar agricultural substances.

Dr. Jones was the first president of the Soil and Plant Analysis Council and then served as its secretary–treasurer for a number of years. He established two international scientific journals, *Communications in Soil Science* and *Plant Analysis and the Journal of Plant Nutrition*, serving as executive editor of each during the early years of their publication.

Dr. Jones is considered an authority on applied plant physiology and the use of analytical methods for assessing the nutrient element status of rooting media and plants as a means for ensuring plant nutrient element sufficiency in both soil and soilless crop production settings.

The author currently lives in Anderson, South Carolina and can be contacted by mail at GroSystems, Inc., 109 Concord Road, Anderson, SC 29621 and by email at: jbhydro@carol.net.

# Section I

## Introduction and Basic Principles

# 1 Introduction

Successful crop production requires knowledge and skill on the part of the farmer/ grower when preparing the soil, selecting inputs, planting the crop, and then managing the crop from emergence to harvest. Knowledge has two sources: that obtained from reliable sources and that learned from the hard knocks of experience. Even the most knowledgeable need at times to refer to a reliable source for refreshing their memory or to learn what is new. The basic principles of soil fertility and plant nutrition are fairly well established. It is the application of these principles that is constantly changing as procedural practices adapt to new products, systems of crop management procedures, and plant genetics.

## 1.1 MANAGEMENT REQUIREMENTS

Management requirements for achieving a moderate yield and average product quality require fewer inputs and skill requirements than are required for achieving maximum yield and highest product quality, the latter not allowing for errors in procedural practices. For most cropping situations, maximum biological yield potential based on the combination of soil and plant parameters is not known. It is also not possible to advance quickly from a moderate soil fertility/plant nutrition status to one that results in high yield/quality product achievement. Those management practices applied to one set of soil/plant/climatic conditions are not applicable to all ranges of conditions.

## 1.2 PRODUCTIVITY FACTORS

Some of the most productive soil/plant/climatic areas in the world consist of a unique combination of these three characteristics. For example, the productive soil/plant/ climatic valleys in southern California are not repeated in many other regions of the world where similar crops are grown. The author compared the yield and quality of vegetable crops grown in the valley areas of California with the same crops grown in southern Georgia. The major factor contributing to quality is nighttime summer air temperature—cool in the California valleys, frequently hot and humid in southern Georgia. High corn grain yields can be achieved under a fairly wide range of soil/ climatic conditions, from the dry irrigated fields of central Nebraska, to the rain-fed fields of central Iowa, to the irrigated sandy soils of the southeastern Coastal Plain.

## 1.3 CLIMATIC FACTORS

Air temperature, rainfall pattern, wind, day length, and solar radiation intensity contribute to both low and high yield/quality outcomes. It is sometimes the interacting of these five factors with the soil/plant characteristics that determines the outcome,

3

making an evaluation at the end of the growing season difficult unless these climatic factors are taken into account. A corn farmer in southeastern Georgia won the 200-bushel (bu) club state championship even though he did not irrigate his corn crop, and in that year drought conditions kept corn yields low throughout the state. What happened? The farmer explained that almost every afternoon, a dark cloud appeared over his cornfield and a light rain fell, just enough to keep the plants from wilting. In many desert areas of the world, dew is a major source of water, sufficient to satisfy the minimum plant requirement to sustain growth and yield.

## 1.4   MOVING UP THE YIELD SCALE

A corn farmer in central Indiana, dissatisfied with his grain yields, sought out the latest information related to high grain yield production available at that time from research and extension agronomists, soil fertility specialists, and farm advisors. He changed his soil fertility management procedures, tillage and cultural practices, as well as corn variety, plant spacing, plant population, and date of planting. He monitored the crop and made grain yield determinations from selected areas of the field. After each crop year, he made an evaluation of each input as to its contribution to the final grain yield, adjusting those practices that failed to contribute to yield or that could be modified to increase the grain yield potential. After 5 years, his 100-acre grain yields began to exceed 200 bu, a record of considerable accomplishment based on the fact that at that time, 100 to 120 bu per acre yields were considered expected maximum yields using currently recommended management practices.

During this time period, 100-bu corn grain club programs were initiated by either county or statewide cooperative extension programs, designed to assist farmers in achieving grain yields higher than the current state average. Before too long, due to the success of these programs, 100-bu grain yields were being achieved by many farmers, so the goal was increased either to 150- or 200-bu club goals. Today, corn grain yields in excess of 200 bu are common, many due to these programs that assisted farmers by overcoming yield-limiting factors. Similar programs for other crops have resulted in guiding farmers and growers toward more efficient utilization of inputs, diagnosing and eliminating those factors that are yield depressing and applying those practices that will result in higher production and quality of produced product.

## 1.5   PRODUCT QUALITY

The ability to produce a quality product is critical for most crop production systems. Consumers of fruits and vegetables are particularly quality conscious, where physical appearance determines acceptance for purchase or when making a choice of which product is selected. In addition to appearance, factors as to origin, local, regional, out of the country, and methods of production, such as organically produced, can be factors when a choice is made for purchase. The subjects discussed in this book are correlated to product quality by how the principles of soil and plant nutritional management are applied.

# 2 Soil Fertility Principles

The basic soil fertility principles are based on knowledge of the physical and chemical properties of a soil and how these properties impact plant growth. Knowing what these properties are, a soil can be modified by soil manipulation, including both physical procedures and by the application of substances that will alter the existing properties.

## 2.1 FERTILE SOIL DEFINED

What defines a "fertile soil" is determined by the combination of both the physical (texture, structure, profile depth, water-holding capacity, drainage, etc.; see Chapter 7, "Physical Properties of Soils") and physiochemical properties (pH, level of available essential plant elements, cation/anion exchange capacities (see Chapter 8, "Physiochemical Properties of Soil"). A fertile soil may be defined either on the basis of its own physical-chemical properties, or based on crop performance and yield. For example, some compact alkaline desert soils, when fertilized and with adequate water applied, will produce wheat yields at or near world records, while some of the most productive soils in the world, high in organic matter content with a deep soil profile, tilled when wet, will reduce soil tilth, resulting in poor plant growth and product yield.

Other factors, such as mineral (see Chapter 6, "Soil Taxonomy and Horizontal Characteristics") and organic matter content (see Chapter 10, "Soil Organic Matter"), will also contribute to fertility status.

Profile depth and depth to the subsoil can be influencing factors related to crop growth and product yield. Profile depth may limit root growth, resulting in drought conditions when plants are under atmospheric stress with soil water resources limited by soil depth. The pH and fertility status of the subsoil may restrict root growth where water maybe available for plant use. In some instances, there may exist a hard pan at the surface–subsoil interface, a naturally occurring condition, or one created by tillage procedures. Deep plowing to break up a hardpan as well as the introduction of liming material to correct soil acidity and fertilizer to make the subsoil "fertile" can significantly contribute to what would be defined as a "fertile soil."

A "fertile soil" will partially compensate for periods of plant stress occurring as a result of less-than-optimum growing conditions due to air/soil temperature and moisture extremes, extended periods of low light intensity, and long periods of calm or sustained high winds (see Chapter 26, "Weather and Climatic Conditions").

## 2.2  MAKING AND KEEPING A SOIL FERTILE

Most soils are not naturally "fertile," requiring management procedures and treatment in order to establish desired physical and chemical conditions, to be then followed by procedural practices and treatments needed to sustain an established soil fertility level (see Chapter 18, "Liming and Liming Materials," and Chapter 19 "Inorganic Fertilizers and Their Properties"). Some soil properties are not easily changed, or are not changeable without extreme measures. However, there are management procedures that, when correctly employed, will best utilize the existing soil properties while minimizing the effects of those properties that can adversely affect plant growth and crop performance.

What is difficult to define is that soil fertility condition best suited for the cropping routine being followed, particularly for multiple-cropping systems. The essential plant element demands for one crop may impact a following crop, while establishing an "ideal" soil fertility regime for one crop may not be the best for another or other following crops. How a crop is managed will also determine what might occur in following crops. For example, corn grown for grain, leaving the vegetative portion of the plant in the field for soil incorporation, will have less impact on the fertility status of the soil versus corn being grown as a silage crop with the entire upper portion of the plant being harvested, removing essential elements that must be replaced by fertilization.

Some farmers employ "green manure" programs as a means of maintaining soil organic levels as well as providing a means for recirculating essential plant nutrient elements. The planting of a cover crop between growing seasons will minimize soil erosion and the potential loss of essential plant nutrient elements by profile leaching. Turning under a cover crop will provide a source of absorbed essential plant nutrient elements when decomposition occurs.

Adding or plowing under highly carbonaceous materials, such as small grain stover when being decomposed by soil organisms, will draw nitrogen from the available soil pool, thereby reducing that available for a growing crop. Microorganisms are better competitors than plant roots for soil resources. Under such conditions, it may be necessary to apply nitrogen fertilizer sufficient to satisfy the crop requirement as well as that needed for microbial decomposition.

## 2.3  BIOLOGICAL FACTORS

Soil microorganisms play significant roles in defining what determines a fertile soil. Species types and their populations are determined by both soil characteristics, such as pH, organic matter content, temperature, and texture, and the species of the plants growing. Microorganisms require a "food" source that is derived either from plant roots, crop residues, or added organic materials, such as animal manures.

Under natural conditions, there exists in the soil a wide range of microflora, many species at widely varying populations. With cropping, both the range in microorganism species and their population can significantly change. In a mono-culture cropping system, the range in variety of microorganisms decreases and the population of some microorganisms increases. The effect of this change can be seen in what occurs with cropping systems. For example, corn yield in a mono-culture system will be less (10% or more, depending on the soil type and growing conditions) than that obtainable when corn is grown in rotation on the same soil. Both plant stand and bean yield can be significantly less for continuously cropped soybean compared to soybean in rotation. To reinvigorate an already existing alfalfa stand by reseeding may result in slow seeding growth, or even failure of seedling establishment due to the existence of high populations of certain micro-organisms that exhibit a pathological effect on the new alfalfa seedlings. Keeping the microorganism variety range high and populations among the microorganisms in balance is obtained by rotating crops or, to some limited degree, by using cover crops between each monoculture planting.

## 2.4   AN IDEAL SOIL

An ideal soil is characterized as one with

- A loamy texture for ease of air and water movement into the soil
- An organic matter content sufficient to sustain microorganism populations
- Textural and organic matter characteristics that contribute to soil tilth
- A soil structure that promotes proliferation of plant roots into the soil mass, and ease of water drainage and air exchange at the soil surface
- Sufficient clay (as well as organic) colloids to hold reserve essential plant nutrient elements and soil moisture
- A deep soil profile with a permeable subsoil allowing for root penetration and normal soil water drainage
- A subsoil fertility (pH and level of essential plant nutrient elements) that promotes root growth

Some of the procedures needed to establish and sustain a fertile soil are given by Parnes (1990).

## 2.5   SOIL FERTILITY MANAGEMENT CONCEPTS

Soil fertility management has two requirements: establishing and maintaining the soil pH and essential plant nutrient elemental content within their desired ranges for that soil type and crop or cropping sequence employed with its associated cultural management practices. It is obvious that no one soil fertility management system will meet all these requirements; however, there are basic principles that do apply, requiring moderate modification to suit the specifics of soil type, crop species, and climatic/weather characteristics. These influencing factors are discussed in some detail in the various chapters of this book.

There are two concepts for establishing and maintaining the fertility status of a soil:

1. Establishment of a certain soil fertility level, and then by means of soil test tracking (see page 125), treat the soil on the basis of what is needed to maintain that status, relying on the established soil fertility level to carry the crop(s) from planting to harvest without the occurrence of a plant nutrient element insufficiency.
2. Correct soil elemental insufficiencies determined by means of a soil test (see Chapter 16, "Soil Testing") and/or plant analysis (see Chapter 17, "Plant Analysis") of the previous crop; then add what is needed to meet the essential plant nutrient element requirements of the crop from emergence to fruit production.

Using either concept of soil fertility management, there are two challenges:

1. What is required to establish and then maintain an optimum soil fertility condition, including soil pH
2. What defines the plant nutrient elemental status of a soil as being "optimum"

Another fertilization strategy is to apply fertilizer sufficient for the crop to be initially grown with an expected carryover sufficient to meet the needs of a following crop. An example would be a soybean crop following a corn or small grain crop. The danger here is that there may be "luxury consumption" (see page 240) by the first crop with the carryover being less than anticipated. In such a crop sequence procedure, it is assumed that the corn stover will "trap" the essential plant nutrient elements, and with their release on decomposition, be sufficient to meet the needs of the following soybean crop. If decomposition is impaired due to the lack of sufficient soil incorporation and/or weather conditions not conducive for decomposition (mainly soil temperature and moisture), the essential plant nutrient elements in the corn stover will remain "trapped," and therefore not available for use by the soybean plants.

Soil fertility related to soil pH is discussed in Chapter 9, "Soil pH: Its Determination and Interpretation," and the requirements for maintaining the soil pH within a desired range discussed in Chapter 18, "Liming and Liming Materials." Soil pH maintenance is probably the most overlooked soil fertility factor by assuming that a soil kept within a certain pH range will ensure sufficiency in terms of elemental plant nutrient availability (and toxicity), which will then impact plant growth and crop performance, and final plant and/or grain (product) yield. Soil pH maintenance may require liming procedures that mimic that for fertilizing a soil to maintain essential plant nutrient element sufficiency.

## 2.6  MULTIPLE FACTOR YIELD INFLUENCE

The "Law of the Minimum," which has been widely accepted, states "that the final product yield is determined by that factor most limiting." This is frequently illustrated by what determines the water level in a barrel, being that of the shortest stave (Figure 2.1). This concept is erroneous because the final product yield is a multiple

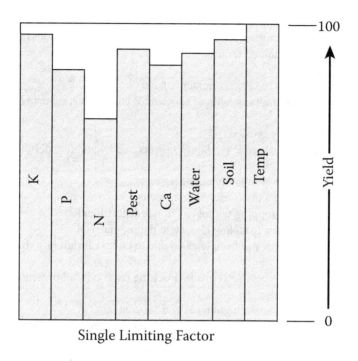

Single Limiting Factor

**FIGURE 2.1**    An illustration of the "Law of the Minimum," that the final yield is deter-mined by the most limiting factor.

of influencing factors. Therefore, if all the factors are at the 100% sufficiency level, then the final yield will be 100. However, if there are five influencing factors, then the final yield will be the multiple of all those factors. For example, if the sufficiency level for each factor is 90%, then the final yield will be 90 × 90 × 90 × 90 × 90 = 56% of the maximum, and not 90%. This explains why yield performance may be considerably less than expected by failing to realize the impact that the multiple fac-tor concept has on yield determination. Naturally, this is a theoretical example, so the interacting impact may not always be as shown in this illustration, with the final yield being either higher or lower.

## 2.7    SOIL CONDITION RELATED TO DEFICIENCY IN A MAJOR ELEMENT AND MICRONUTRIENT

Certain soil characteristics have been associated with the occurrence of major ele-ment and micronutrient deficiencies.

### 2.7.1    MAJOR ELEMENTS

Nitrogen (N):
*   Sandy soils that have been leached by heavy rainfall or irrigation

- Mineral soils low in organic matter content
- Long history of crop-depleting N supply when applied N is less than that required by the planted crop(s)

Phosphorus (P):

- Mineral soils low in organic matter content
- Long history of cropping without adequate P fertilization reducing the supply of P
- P-rich soils lost by erosion
- Calcareous soils where P availability is reduced by alkaline pH

Potassium (K):

- Mineral soils low in organic matter content
- Soils having a low cation-exchange capacity
- Long history of cropping without adequate K fertilization
- Sandy soils formed from low K-content parent material
- Sandy soils when K has been leached due to either rainfall or irrigation

Calcium (Ca):

- Acid sandy soils when Ca is lost by leaching from rainfall or irrigation
- Strongly acid peats
- Alkaline or sodic soils, high in pH and Na content
- Soils with high soluble Al, low exchangeable Ca content

Magnesium (Mg):

- Acid sandy soils when Mg is lost by leaching from rainfall or irrigation
- Acid soils with pH less than 5.4
- Strongly acid peat and muck soils
- Soils over-fertilized with either Ca and/or K

Sulfur (S):

- Mineral soils low in organic matter
- Soils after years of cropping
- Acid sandy soils where sulfates have been leached by rainfall
- Soils formed from low S-containing parent material
- No substantial deposition of S by acid rainfall
- Use of low-S containing NPK fertilizers [i.e., substituting triple superphosphate (0-46-0) for superphosphate (0-20-0)]

## 2.7.2 MICRONUTRIENTS

Boron (B):

- Acid igneous soils
- Sandy soils where B has been leached by either rainfall or irrigation
- Calcareous soils
- Soils low in organic matter
- Acid peat and muck soils

Copper (Cu):

- Peat and muck soils
- Calcareous sands

- Leached acid soils
- Soils formed from low Cu-containing parent materials

Iron (Fe):
- Calcareous soils where available (soluble) Fe is low
- Waterlogged soils
- Acid soils with excessively high soluble Mn, Zn. Cu, and Ni contents
- Sandy soils low in total Fe
- Peat and muck soils

Manganese (Mn):
- Calcareous soils where Mn availability is low
- Poorly drained soils high in organic matter content
- Strongly acid sandy soils where Mn has been leached by either rainfall or irrigation
- Soil formed from low Mn-content parent materials

Molybdenum (Mo):
- Low in sandy soils
- Continuously increases in availability with increasing pH
- Liming frequently corrects an Mo deficiency with some crops

Zinc (Zn):
- Alkaline soils
- Sandy soils leached by either rainfall or irrigation
- Leveled soils where Zn-deficient subsoils are exposed on the surface
- Soils where heavy, frequent applications of P have been applied

## 2.8  ELEMENTAL CONTENT OF THE SOIL AND SOIL SOLUTION

A soil has both a solid and a liquid phase. The essential plant mineral nutrient elements exist in four solid forms:

1. In minerals that are water insoluble
2. In minerals that are slightly water soluble
3. As ions held on the exchange sites of soil colloids
4. As a constituent in soil organic matter

The release of elements from the solid phase into the soil solution is the result of an ever-changing complexity of the dynamic chemical and biological activities occurring in the soil. The rate at which this process occurs depends on a number of soil factors:

- pH
- Soil moisture content
- Physiochemical characteristics of the colloidal substances
- Solubility characteristics of the solid-phase components
- Temperature
- Biological activity

Elements released from the solid phase into the liquid phase, called the soil solution, exist as ions. An element in its ionic form must be present in the soil solution in order to be absorbed by plant roots. Elemental root absorption does not occur for an element adsorbed onto the surface of a soil particle even though there is direct physical contact between a soil particle and root surface. Absorption only occurs from the soil solution. The concentrations of ions in the liquid phase are in equilibrium with those in the various four solid-phase forms listed above. As an element ion is absorbed by the plant root from the soil solution, the equilibrium shifts, resupplying the soil solution and thereby maintaining the equilibrium.

How these ions are brought into proximity to the root occurs by means of three processes: mass flow, diffusion, and root interception.

1. *Mass flow* occurs when water moves within the soil mass, carrying dissolved ions along with the moving water. For example, the ions of Ca ($Ca^{2+}$) and N [as the nitrate ($NO_3^-$) anion] are primarily moved in the soil by mass flow. These ions can be carried considerable distances by this process. However, if the soil moisture content is low, movement by mass flow will be impaired. In addition, water draining from the soil will also carry dissolved elements out of the rooting zone. Utrafication of streams and lakes and the accumulation of elements in groundwater can be linked to that coming from elements released by mineralization, organic matter decomposition, or from added fertilizer amendments. Water movement can occur in three directions: down through the soil profile as a result of rainfall or applied irrigation water, pulled up through the soil profile by the evaporation of water at the soil surface, and to some degree laterally within the soil profile from an advancing water front.

2. *Diffusion* is the process by which ions move within the soil solution from an area of high concentration to an area of lower concentration. Most element ions (see page xviii) move by diffusion in the soil solution surrounding plant roots. As ions are root absorbed from the soil solution, a concentration gradient is created that moves ions from surrounding areas of higher concentration to this lower concentration area at the root interface. Movement by this process is measured in a few millimeters. It takes a very low soil moisture condition in order to effect ion movement by diffusion.

3. *Root interception* occurs as plant roots expand into the soil mass, resulting in an ever-increasing root surface contact with soil particles and their surrounding soil solution. Root exploration can be both beneficial to the plant by increasing contact with fertile soil, and also detrimental as roots venture into soil areas of low or high pH, devoid of essential plant elements, or soils that have high "available" levels of elements that can be toxic to plant roots as well as the plant itself (see Chapter 14, "Elements Considered Toxic to Plants").

It should be remembered that even with an extensive plant root system, very little of the total soil mass is in immediate contact with plant roots. Therefore, the importance of mass flow and diffusion that bridges the gap existing between soil particles and root surfaces. Both are necessary functioning processes, ensuring essential plant nutrient element sufficiency for the growing plant. With any one of the three processes, mass flow, diffusion, and root interception, impaired, essential plant nutrient element deficiencies are likely to occur.

# 3  Plant Nutrition Principles

The use of the word "nutrition" can be confusing, as *plant nutrition* is a broad term that would apply to all aspects of plant growth. *Plant mineral nutrition* would relate to just the elements identified as minerals whose presence or absence could affect the growth of plants. Even the word *mineral* can be misleading as it has the connotation of being a compound of elements. Another word that has crept into the plant nutrition jargon is *metal*, which would refer to those elements that are identified as metals, such as Fe, Cu, Mn, and Zn. The other word that can be misunderstood is *nutrient,* as it does not specifically have the connotation as just being an element or mineral. In some instances, both nutrient and element are combined in defining those elements that are known as essential to be a nutrient element. Therefore in this chapter, *plant nutrition* is defined as the study of those elements that are essential for plants to grow, and the combination of words, essential plant nutrient element, will be used to identify those elements essential to plants.

In the Wikipedia definition (www.Wikipedia.org) of plant nutrition, fourteen elements are given as essential nutrients (the author would choose the word "element" in place of "nutrient"), to include the element Ni, an element that has not been widely accepted as being essential, although its identification as a micronutrient is becoming commonplace in both the technical and scientific literature (see Chapter 13, "Elements Considered Beneficial to Plants"). The three elements C, H, and O are not considered plant nutrient elements in the Wikipedia definition. The author classifies these three as "structural elements" because they are the primary elements comprising those substances in plants that form the plant skeleton (cell walls, conductive tissue, etc.).

There are several principles that apply to the subject of plant nutrition. Some plant nutrient elements are directly involved in plant metabolism, while others are part of the cellular structure of the plant. A plant nutrient element that is able to limit plant growth according to Liebig's Law of the Minimum (see page 8) is considered an essential plant nutrient element if the plant cannot complete its life cycle without it. Plants require specific element concentrations during vegetative growth, flowering, and fruit production.

## 3.1  PHOTOSYNTHESIS

Without green plants, whose leaves contain chlorophyll, our planet would be a very barren place. In the process called *photosynthesis*, chlorophyll (Figure 3.1), when exposed to sunlight (wavelengths between 400 and 700 nm visible light), is able to convert photon energy into chemical energy (plant carbohydrates). By splitting a water ($H_2O$) molecule and combining the hydrogen proton ($H^+$) with a carbon dioxide

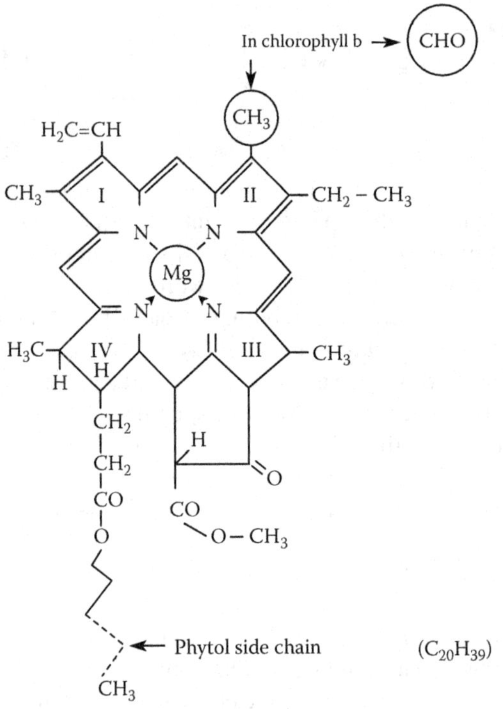

In chlorophyll b → CHO

$H_2C=CH$

$CH_3$

$CH_3$ — I   II — $CH_2$ – $CH_3$

N   N

Mg

N   N

$H_3C$ — IV   III — $CH_3$

H   H

$CH_2$   H

$CH_2$   O

CO   CO

O   O – $CH_3$

← Phytol side chain          $(C_{20}H_{39})$

$CH_3$

**FIGURE 3.1**   Molecular structure of the chlorophyll molecule.

($CO_2$) molecule, a carbohydrate molecule is formed and a molecule of oxygen ($O_2$) is released, as is illustrated in the following equation:

Carbon dioxide ($6CO_2$) + Water ($6H_2O$)½

↓

(in the presence of light and chlorophyll)

↓

Carbohydrate ($C_6H_{12}O_6$) + Oxygen ($6O_2$)

In the photosynthesis process, there are two biochemical reactions that lead to the production of carbohydrates, one that occurs in C3 plant species, the other in C4; the 3 and the 4 designate the number of C atoms that exist in the first product of photosynthesis. For those plant species designated as C3 (see page 237), their photochemical process follows what is known as the Calvin cycle, named after the man who isolated the first product of synthesis. Dr. Calvin was awarded a Nobel Prize for his discovery. There are a number of significant differences between C3 and C4 (see page 237) plant species, one being cellular leaf structure differences that affect $CO_2$ fixation, with C4 plant species more efficient in their absorption of $CO_2$. C3 plant species are more responsive to the $CO_2$ content in the air. The association between air temperature and $CO_2$ air content is less an influencing factor for C3 than C4 plant species. C4 plant species have a higher water-use efficiency, grow well in hot environments, have a higher productivity potential and optimum air temperature requirement, and lower transpiration potential and photorespiration rates than C3 plant species. C4 plant species do not grow well in low light environments. Most of the world's (~300,000) plant species are C3. The major C4 food plants are corn, millet, sorghum, and sugar cane and are grown worldwide, while many of the C3 food plant species have specific adaptation requirements.

## 3.2 THE FUNCTION OF PLANTS

In addition to providing food and fiber, plant activity also

- Maintains the balance of atmospheric oxygen ($O_2$) and carbon dioxide ($CO_2$)
- Is a major source of atmospheric moisture through transpiration
- Controls soil erosion
- Recycles soil elements
- Is a source of beauty and wonder because of the wide range of growing and flowering habits, producing a rainbow of foliage and flower colors

Although there are many thousand differing kinds of plants, relatively few species are grown as a source of food for human use. Grain crops (corn, wheat, and rice) and the root crops (potato and cassava) provide much of the carbohydrate in human diets, while fruits and vegetables are the major sources of dietary protein and vitamins.

Cotton is the main fiber crop for making cloth, while trees provide the source for building materials, paper, and fuel. Plants are also the sources for a number of important industrial chemicals and pharmaceuticals.

Plants are far-ranging in their growth habits, cellular complexity, reproduction characteristics, and requirements for growth such as temperature and moisture tolerance, response to changing light conditions, and essential plant nutrient element requirements. Most plants have either a wide or narrow range of adaptability to these environmental conditions.

The nutritional requirements of plants also vary, a factor that is being effectively manipulated by humans to alter both yield and quality. In addition, plants have been genetically modified to enhance the environment and extend their utilization.

A wide range of plants, trees, shrubs, and flowering perennials and annuals are used in indoor landscapes to enhance the beauty of buildings and homes. Various grass species have been selected and bred for use as turf on athletic fields, golf courses (tees, fairways, and greens) as well as for commercial and home lawns.

Plants are being studied for their adaptation to space enterprises, serving as a means of absorbing $CO_2$, supplying $O_2$, and recycling water and human wastes as well as providing a potential source of food.

## 3.3 DETERMINATION OF ESSENTIALITY

It was not until the 1800s that scientists began to unravel the mysteries of how green plants grow. A number of theories were put forth to explain plant growth, but through observation and carefully crafted experiments, scientists began to learn what were the essential requirements for normal growth and development.

By the beginning of the 1900s, ten of the sixteen elements now known as required by plants had been identified.

It might be worthy to note that the various humus-concept theories relating elemental form to plant "health" and growth had their origins in theories developed by some of these early scientists. The idea that the soil provided food for plants, or that the humus in the soil was the source of plant health, still has its proponents today. It has been fairly well established that the *form* of an essential plant nutrient element, whether as an inorganic ion or having its origin derived from an organic matrix, is not a factor that determines the well-being of the plant. From a mineral nutrition standpoint, it is the combination of concentration and labile form of an essential plant nutrient element that determines the elemental status of a plant.

Those early scientists had also discovered that the mass of a live plant was essentially composed of water and organic substances, and that in most plants the mineral matter constituted less than 10%, and frequently less than 5%, of the dry matter content of the plant. From the analysis of the ash, after removal of water and the destruction of the organic matter, scientists began to better understand the elemental requirements of plants, noting which elements were present in the ash and at what concentrations. However, at that time there was no system for scientifically establishing the absolute essentiality of elements found in the ash; just their presence was assumed to be related to their essentiality.

By 1890, scientists had already established that the elements N, P, S, K, Ca, Mg, and Fe were required by plants, and that their absence or low availability resulted in either the death of the plant or very poor plant growth with accompanying visual symptoms of growth abnormalities. Between 1922 and 1954, additional elements were determined to be essential, those elements being Mn, Cu, Zn, Mo, B, and Cl. It is not surprising that many of the essential plant nutrient elements were not identified until the purification of reagent chemicals was achieved, and the techniques of analytical chemistry had brought detection limits to below the milligram level.

In 1939, two plant physiologists at the University of California published their criteria for plant nutrient element essentiality, criteria that are still acknowledged today. Arnon and Stout (1939) established three criteria for essentiality:

1. Omission of the element in question must result in abnormal growth, failure to complete the life cycle, or premature death of the plant.
2. The element must be specific and not replaceable by another.
3. The element must exert its effect directly on growth or metabolism and not some indirect effect such as by antagonizing another element present at a toxic level.

For some, these criteria for essentiality are too restrictive, stifling the search of additional essential elements. Neilson (1984) has suggested the following as criteria for essentiality:

"An element shall be considered essential for plant life if a reduction in tissue concentration of the element below a certain limit results consistently and reproducibly in an impairment of physiologically important functions and if restitution of the substance under otherwise identical conditions prevents the impairment, and the severity of the signs of deficiency increases in proportion of the reduction of exposure to the substance."

Plant physiologists of today are still attempting to determine if there are additional elements that are essential to plants, applying the three requirements of essentiality as set forth by Arnon and Stout (1939) more than 70 years ago. The more recent suggestion by Neilson (1984) as criteria for "essentiality" has yet to impact our current concepts. Plant physiologists are still actively engaged in determining what additional elements can be added to the current list of sixteen (see Chapter 13, "Elements Considered Beneficial to Plants").

## 3.4 ESSENTIAL ELEMENT CONTENT IN PLANTS

The concentration of essential plant nutrient elements in the plant required for normal growth and development varies considerably (the relative range being from 1 to 1 million) among thirteen of the essential elements as is shown in Table 3.1 for the sixteen essential elements by characteristics in Table 3.2 and approximate concentration for the ten major elements in Table 3.3.

For diagnostic purposes, the content of the essential plant nutrient elements are given Chapter 17, "Plant Analysis."

The soil factors that affect the elemental concentrations in plants include

**TABLE 3.1**

**Average Concentration of Mineral Elements in Plant Dry Matter Sufficient for Adequate Growth**

| Element | mmol/g | mg/kg (ppm) | % | Relative Number of Atoms |
|---|---|---|---|---|
| Molybdenum (Mo) | 0.001 | 0.1 | — | 1 |
| Copper (Cu) | 0.10 | 6 | — | 100 |
| Zinc (Zn) | 0.30 | 20 | — | 300 |
| Manganese (Mn) | 1.0 | 50 | — | 1,000 |
| Iron (Fe) | 2.0 | 100 | — | 2.000 |
| Boron (B) | 2.0 | 20 | — | 2,000 |
| Chlorine (Cl) | 3.0 | 100 | — | 3,000 |
| Magnesium (Mg) | 80 | — | 0.2 | 80,000 |
| Phosphorus (P) | 60 | — | 0.2 | 60,000 |
| Calcium (Ca) | 125 | — | 0.5 | 125,000 |
| Potassium (K) | 250 | — | 1.0 | 250,000 |
| Nitrogen (N) | 1,000 | — | 1.5 | 1,000,000 |

*Source:* Epstein, E. 1965. Mineral nutrition, pp. 438–466. In J. Bonner and J.E. Varner (Eds.), *Plant Biochemistry*. Academic Press, Orlando, FL.

**TABLE 3.2**

**Characteristics of the Nutrient Elements Essential for Plant Growth, Their Principal Form for Uptake, and Plant Content**

| Element | Atomic Number | Atomic Weight | Principle Forms of Uptake | Plant Content % mole/g | Range, % |
|---|---|---|---|---|---|
| **Macronutrients** | | | | | |
| Hydrogen (H) | 1 | 1 | Water | 60,000 | |
| Carbon (C) | 6 | 12 | Air ($CO_2$) | 40,000 | |
| Oxygen (O) | 8 | 16 | (soil) | | |
| | | | Water $H_2O$ | 30,000 | |
| Nitrogen (N) | 7 | 14 | $NH_4^+$, $NO_3^-$ | 1,000 | 0.5–5.0 |
| Potassium (K) | 19 | 39.1 | $K^+$ | 250 | 0.5–5.0 |
| Calcium (Ca) | 20 | 40.1 | $Ca^{2+}$ | 125 | 0.05–5.0 |
| Magnesium (Mg) | 12 | 24.3 | $Mn^{2+}$ | 80 | 0.1–1.0 |
| Phosphorus (P) | 15 | 31 | $H_2PO_4^-$, $HPO_4^{2-}$ | 60 | 0.1–0.5 |
| Sulfur (S) | 16 | 32 | $SO_4^{2-}$ | 30 | 0.05–0.5 |
| **Micronutrients** | | | | | |
| | | | | **ppm** | |
| Chlorine (Cl) | 17 | 35.5 | $Cl^-$ | 3 | 100–10,000 |
| Boron (B) | 5 | 10.8 | $H_3BO_3$ | 2 | 2–100 |
| Iron (Fe) | 26 | 55.9 | $Fe^{2+}$. $Fe^{3+}$ | 2 | 50–1,000 |
| Manganese (Mn) | 25 | 54.9 | $Mn^{2+}$ | 1 | 20–200 |
| Zinc (Zn) | 30 | 65.4 | $Zn^{2+}$ | 0.3 | 10–100 |
| Copper (Cu) | 29 | 63.5 | $Cu^{2+}$ | 0.1 | 2–20 |
| Molybdenum (Mo) | 42 | 96.0 | $MoO_4^{2-}$ | 0.001 | 0.1–10 |

- Soil test level
- Soil moisture movement of ions, affects K and Mg
- Temperature (affected elements, N, P, K, S, Mg, B, and Zn) decomposition of organic matter
- Soil pH (low pH, increases Mn, Fe, and Al uptake, lowers Mg and P; high pH decreases Fe, Al, Mn, Zn, and B, increases Mo)
- Tillage and placement
- Compaction

The plant factors that affect the elemental concentrations in plants include

- Hybrid or variety
- Stage of growth
- Interaction among the elements, such as P and Zn; P and Mn; K, and Ca, Mg, etc.

**TABLE 3.3**

**Approximate Concentrations of Essential Plant Nutrient Elements Required for Healthy Plant Growth**

| Element | Concentration in Dry Matter | |
| --- | --- | --- |
| | ppm | % |
| Hydrogen (H) | 60,000 | 6 |
| Carbon (C) | 420,000 | 42 |
| Oxygen (O) | 480,000 | 48 |
| Nitrogen (N) | 14,000 | 1.4 |
| Potassium (K) | 10,000 | 1.0 |
| Calcium (Ca) | 5,000 | 0.5 |
| Magnesium (Mg) | 2,000 | 0.2 |
| Phosphorus (P) | 2,000 | 0.2 |
| Sulfur (S) | 1,000 | 0.1 |
| Chlorine (Cl) | 100 | |
| Iron (Fe) | 100 | |
| Boron (B) | 20 | |
| Manganese (Mn) | 50 | |
| Zinc (Zn) | 20 | |
| Copper (Cu) | 6 | |
| Molybdenum (Mo) | 0.1 | |

*Source:* Grunden, N.J. 1987. *Hungry Crops: A Guide to Nutrient Element Deficiencies in Field Crops.* Department of Primary Industries, Queensland Government Publication, Brisbane, Australia.

## 3.5 CLASSIFICATION OF THE THIRTEEN ESSENTIAL MINERAL ELEMENTS

Thirteen of the essential mineral elements have been divided into two categories, based entirely on that concentration needed in the plant in order for them to carryout their functions. Those elements at the highest concentration requirement (as a percent of the dry weight) are termed the major elements, N, P, K, Ca, Mg, and S (see Chapter 11, "Major Essential Plant Elements"). Boron, Cl, Cu, Fe, Mn, Mo, and Zn have lower plant concentration requirements (as a fraction of the dry weight) and are called micronutrients. (see Chapter 12, "Micronutrients Considered Essential to Plants"). The books by Epstein and Bloom (2005), Glass (1989), Marshner (1995), and Mengel and Kirby (1987) are the major texts on plant mineral nutrition.

## 3.6 ROLE OF THE ESSENTIAL PLANT NUTRIENT ELEMENTS

A summarization of the roles of the essential plant nutrient elements is given in Table 3.4, and a more detailed description of function is given in Table 3.5.

**TABLE 3.4**
**Essential Plant Nutrient Elements by Form Utilized and Their Biochemical Function**

| Essential Element | Form Utilized | Biochemical Functions |
|---|---|---|
| C, H, O | $CO_2$, $H_2O$ | Are combined in the photosynthesis process to form a carbohydrate that becomes the physical structure of the plant |
| N, S | $NO_3^-$, $NH_4^+$, $SO_4^{2-}$ | Combine with carbohydrates to form amino acids and proteins that become involved in enzymatic processes |
| P | $PO_4^{3-}$, $H_2PO_4^-$, $HPO_4^{2-}$ | Involved in the energy transfer reactions |
| B | $H_3BO_3$, $BO_3^{3-}$ | Involved in carbohydrate reactions |
| K, Mg, Ca, Cl | $K^+$, $Ca^{2+}$, $Mg^{2+}$, $Cl^-$ | Involved in the osmotic potentials, balancing anions, controlling membrane permeability and electropotentials |
| Cu, Fe, Mn, Zn, Mo | $Cu^{2+}$, $Fe^{2+}/Fe^{3+}$, $Zn^{2+}$, $MoO_4^{2-}$ | Enable electron transport by valency change |

## 3.7   PLANT NUTRIENT ELEMENT SOURCES

Other than C, H, and O, the essential plant nutrient elements are primarily taken up through the roots as ions that exist in the soil solution.

Elements can exist in the soil in either inorganic or organic forms, or both. Inorganic sources are soil minerals and that added to the soil by liming (see Chapter 18, "Liming and Liming Materials") or the addition of fertilizers—either inorganic (see Chapter 19, "Inorganic Fertilizers and Their Properties") or organic (see Chapter 20, "Organic Chemical Fertilizers and Their Properties"). Organic debris, plant residues, and microorganisms are the major sources for the elements B, N, P, and S. As plant and microorganism residues decay, ions of these elements are released into the soil solution. The rate and extent of decomposition that results in their release depend on soil temperature, moisture, and degree of aeration.

The soil and the soil solution are components of an ever-changing complex of dynamic chemical and biological systems that are determined by the soil's physical (see Chapter 7, "Physical Properties of Soil") and physiochemical (see Chapter 8, "Physiochemical Properties of Soil") properties as well as being influenced by soil temperature, moisture content, pH, level of elements present [whether essential (see Chapter 11, "Major Essential Plant Elements," and Chapter 12, "Micronutrients Considered Essential to Plants"), beneficial (see Chapter 13, "Elements Considered Beneficial to Plants"), or toxic (see Chapter 14, Elements Considered Toxic to Plants")], and degree of aeration (see Chapters 2 and 4). An element in its ionic form must exist in the soil solution in order to be absorbed by plant roots. How these ions are brought into proximity to the roots has been categorized by three processes: mass flow, diffusion, and root interception (see Chapter 2).

## TABLE 3.5
## Plant Functions of the Essential Elements

### Major Elements

*Nitrogen (N):*
- Found in both inorganic and organic forms in the plant
- Combines with C, H, O, and sometimes S, to form amino acids, amino enzymes, nucleic acids, chlorophyll, alkaloids, and purine bases
- Organic N predominates as high-molecular-weight proteins in plants
- Inorganic N can accumulate in the plant, primarily in stems and conductive tissue, in the nitrate ($NO_3$) form

*Phosphorus (P):*
- A component of certain enzymes and proteins, adenosine triphosphate (ATP), ribonucleic acids (RNA), deoxyribonucleic acids (DNA), and phytin
- ATP is involved in various energy transfer reactions, and RNA and DNA are components of genetic information

*Potassium (K):*
- Involved in maintaining the water status of the plant, the turgor pressure of its cells, and the opening and closing of its stomata
- Required for the accumulation and translocation of newly formed carbohydrates

*Calcium (Ca):*
- Plays an important part in maintaining cell integrity and membrane permeability; enhances pollen germination and growth
- Activates a number of enzymes for cell mitosis, division, and elongation
- May also be important for protein synthesis and carbohydrate transfer
- Its presence may serve to detoxify the presence of heavy metals in the plant

*Magnesium (Mg):*
- A component of the chlorophyll molecule (see Figure 3.1)
- Serves as a cofactor in most enzymes that activate phosphorylation processes as a bridge between pyrophosphate structures of ATP or ADP and the enzyme molecule
- Stabilizes the ribosome particles in the configuration for protein synthesis

*Sulfur (S):*
- Involved in protein synthesis
- Is part of the amino acids cystine and thiamine
- Is present in peptide glutathione, coenzyme A, and vitamin $B_1$, and in glucosides, such as mustard oil and thiols that contribute the characteristic odor and taste to plants in the *Cruciferae* and *Liliaceae* families
- Reduces the incidence of disease in many plants

### Micronutrients

*Boron (B):*
- Believed to be important in the synthesis of one of the bases for RNA (uracil) formation
- In cellular activities (i.e., division, differentiation, maturation, respiration, growth, etc.)
- Long been associated with pollen germination and growth, improving the stability of pollen tubes
- Relatively immobile in plants
- Transported primarily in the xylem

*(continued)*

**TABLE 3.5** (*continued*)
**Plant Functions of the Essential Elements**

### Micronutrients

*Chlorine (Cl):*
- Involved in the evolution of oxygen ($O_2$) in photosystem II in the photosynthetic process
- Raises the cell osmotic pressure
- Affects stomatal regulation
- Increases the hydration of plant tissue
- May be related to the suppression of leaf spot disease in wheat and fungus root disease in oat

*Copper (Cu):*
- Constituent of the chloroplast protein plastocyanin
- Serves as part of the electron transport system linking photosystems I and II in the photosynthetic process
- Participates in protein and carbohydrate metabolism and nitrogen ($N_2$) fixation
- Is part of the enzymes that reduce both atoms of molecular oxygen ($O_2$) (cytochrome oxidase, ascorbic acid oxidase, and polyphenol oxidase)
- Is involved in the desaturation and hydroxylation of fatty acids

*Iron (Fe):*
- Important component in many plant enzyme systems, such as cytochrome oxidase (electron transport) and cytochrome (terminal respiration step)
- Component of protein ferredoxin and is required for $NO_3$ and $SO_4$ reduction, nitrogen ($N_2$) assimilation, and energy (NADP) production; Functions as a catalyst or part of an enzyme system associated with chlorophyll formation
- Thought to be involved in protein synthesis and root-tip meristem growth
- The plant and soil chemistry of Fe is highly complex, with both aspects still under intensive study

*Manganese (Mn):*
- Involved in the oxidation–reduction processes in the photosynthetic electron transport system
- Essential in photosystem II for photolysis, acts as a bridge for ATP and enzyme complex phosphokinase and phosphotransferases, and activates IAA oxidases
- Not known to interfere with the metabolism or uptake of any of the other essential elements

*Molybdenium (Mo):*
- Is a component of two major enzyme systems: nitrogenase and nitrate reductase, nitrogenase being involved in the conversion of nitrate ($NO_3$) to ammonium ($NH_4$)
- The requirement for Mo is reduced greatly if the primary form of nitrogen (N) available to the plant is $NH_4$

*Zinc (Zn):*
- Involved in the same enzymatic functions as Mn and Mg, with only carbonic anhydrase being activated by Zn
- The relationship between P and Zn has been intensively studied as research suggests that high P can interfere with Zn metabolism as well as affect the uptake of Zn through the root
- High Zn can induce an Fe deficiency, particularly those sensitive to Fe

## 3.8   ELEMENT ABSORPTION AND TRANSLOCATION

In general, during rapid vegetative growth and development, the uptake of element ions from the rooting medium is substantial, and as the plant approaches maturity, the rate of accumulation begins to decline. There also exists an N form preference during the early growth of some plants for the ammonium ($NH_4$) cation versus the nitrate ($NO_3$) anion. With time, this preference declines; and as the plant approaches maturity, $NO_3$ can accumulate in the plant at fairly high concentrations if there is a substantial N supply in the rooting medium.

The uptake of element ions and their distribution and redistribution within the plant are governed by time. For example, shortly after germination in soil, the Al concentration (frequently Fe also) found in the plant can be very high, a concentration level that would be considered toxic for the more mature plant, although the newly forming plant seems unaffected. But within a few weeks after germination, the Al content in the plant declines sharply.

During the reproductive (flowering and seed and/or fruit development) period, considerable redistribution of elements accrues, although the rate and extent vary with element. Thus, the plant nutrient elements can be classified by their mobility within the plant from the most mobile to the least mobile:

- Very mobile: Mg, N, P, and K
- Slightly mobile: S
- Immobile: Cu, Fe, Mo, and Zn
- Very immobile: B and Ca

These plant nutrient element mobility characteristics will determine in what portion of the plant one would expect deficiency symptoms to appear, the most mobile occurring in the older leaves, and the least mobile in the newly emerging and young leaves. Fruit disorders that are associated with either B or Ca (blossom-end rot, for example) not only occur because of their inadequate supply, but being very immobile, their movement from other portions of the plant into the developing fruit is minimal.

As the plant matures, due to reduced uptake and the redistribution of the plant nutrient elements, a large change in their concentration occurs in the older and younger portions of the plant. With maturity, for example, N, P, and K content in leaves declines, while the Ca and Mg content increases. These changes are the result of two factors—movement out of the maturing leaves for the mobile elements and a decrease in dry weight (loss of soluble carbohydrates)—thus affecting the relative relationship that exists between plant nutrient element and organic contents.

These relative changes with time become important factors when assaying the plant to determine its plant nutrient element status. Therefore, the time and plant part selected for analysis and evaluation are important considerations when conducting either a plant analysis or tissue test (see Chapter 17, "Plant Analysis and Tissue Testing").

Root uptake of an ion does not mean that the absorbed ion will be the automatically translocated into the other portions of the plant. As with the root, there exists a mechanism of transport that carries ions across cell membranes and on into the vascular system, which is as complex as that required for ions to enter the root (see Chapter 4).

In general, long-distance upward movement of ions from the root to the growing point is through the xylem, a vessel transport system that carries both water and ions. The downward movement in plants occurs in the phloem, which takes place in living cells. The driving force that moves water, ions, and other dissolved solutes in this complex vascular system comes from a number of sources:

- Transpiration of water from the leaf surfaces of the plant, which draws water from the rooting medium into the root and then up the entire plant
- Root pressure exerted from the roots themselves, pushing water and ions up the plant
- Source-sink phenomenon, which draws water, ions, and solutes from inactive to active expanding portions (growing points, developing fruit, grain, etc.) of the plant

The movement of ions, molecules, and solutes in the xylem is determined to a considerable degree by the transpiration rate that, in turn, has an effect on the distribution of these substances into the stems, petioles, leaves, and fruit. In addition, the movement of these various substances is not uniform in terms of rate and type. For example, transpiration enhances the uptake and translocation of uncharged molecules to a greater extent than that of ions.

Both Si and B have been extensively studied, relating transpiration rate with the distribution of these two elements in various plant parts—the higher the transpiration rate by a particular plant part, the higher the concentration of that element in that plant part. It has been found, for example, that the transpiration rate has a considerable effect on the movement of Ca, a lesser effect on Mg, and minimal influence on K into developing fruit.

Solute and ion movement is unidirectional in the xylem, but bidirectional in the phloem—from source to sink. There is also some cross-transfer from the xylem into the phloem, but not from the phloem into the xylem. The transport rate in the xylem can range from 10 to 100 cm per hour, while that in the phloem is considerably less.

There is also a re-translocation of elements from the shoot to the roots, which has a regulating effect on the uptake rate through the roots. For example, about 20% of the root–shoot transport is taken up by K, related in part to its role as a counter ion for nitrate ($NO_3$) transport in the xylem, a requirement needed for maintenance of cation–anion balance.

As can be seen from this discussion, the movement of ions, molecules, solutes, and water occurs within a fairly complex system of vessels and cells, movement that is driven by both external and internal factors. In general, it is the transpiration process that is the main driving force carrying ions from the roots to the upper portions of the plant. The redistribution of substances once within the plant—plus simple and complex carbohydrates, amino acids, and proteins formed by photosynthetic activity—then becomes fairly complex, regulated by many interacting factors.

## 3.9  ELEMENTAL ACCUMULATION

Some plants are element accumulators while others have the ability to exclude some elements. Some of these elements are naturally occurring in the environment, and others have been added to the environment by human activity. There is also another factor: Some plant genotypes have a different ion-selective ability than do other genotypes. An example is the difference in elemental make-up between legumes and grasses. Legumes will have a higher content of Ca and Mg than K; the opposite is true for grasses (higher K in the plant than either Ca or Mg). Some plants are well adapted to specific soil conditions, such as soil salinity, as they are able to cope with the high concentrations of salt [sodium chloride (NaCl)] in the soil solution.

## 3.10  ELEMENT ABSORPTION AND PLANT GENETICS

Some genotypes can more easily absorb Fe from the soil solution (as well as other elements) than other genotypes. This ability to absorb Fe, for example, has led to the classification of some genotypes as *Fe-efficient,* while others are designated *Fe-inefficient.* Iron-efficient genotypes are able to either acidify the rhizosphere, the thin cylinder of root surface and contacting soil, and/or release Fe-fixing or -chelating substances, such as siderophores, which are the most commonly released substances.

Genotype differences are being used in breeding programs to select on the basis of tolerance to certain soil conditions (such as soil salinity) and to remove from the gene pool undesirable traits, such as sensitivity to a particular element or suite of elements, or to reduce the affinity for elements that might be toxic or make the plant unsuitable for use as feed for animals or food for human consumption.

## 3.11  PLANT NITROGEN FIXATION

Leguminous plants can also obtain N by means of symbiotics $N_2$ fixation. Nitrogen-fixing bacteria invade the plant roots of legumes and form a colony that takes shape as a nodule on the root. These bacteria receive their energy as carbohydrates from the plant, and they in turn fix atmospheric $N_2$ into usable N for the plant. Depending on the strain of bacteria, the number of nodules formed, the aerobic status of the soil surface horizon, and the elemental nutritional status of the plant, sufficient N can be fixed to satisfy the N requirement of the plant. However, the presence of nodules on the roots is not sufficient evidence of nodule activity and performance as the ability of the nodule bacteria to fix atmospheric $N_2$ depends on:

- Fertility level of the soil and nutritional status of the plant
- Status of available mineral N in the soil solution
- Strain of bacteria

Cobalt (Co) is required by the $N_2$-fixing bacteria to function normally. The efficiency of $N_2$ fixation is enhanced by a sound soil and plant nutritional status that ensures a normal healthy growing plant.

If readily available N is present in the soil solution, the efficiency of fixation decreases. If the available N supply is high, or a significant quantity of fertilizer N has been applied, nodule formation will be significantly impaired.

## 3.12 DIAGNOSTIC PLANT SYMPTOMS OF ESSENTIAL PLANT NUTRIENT ELEMENT INSUFFICIENCIES

When an essential plant nutrient element insufficiency (deficiency and/or toxicity) occurs, visual symptoms may or may not appear, although normal plant development will be slowed. When visual symptoms do occur, such symptoms can frequently be used to identify the source of the insufficiency. Visual symptoms of deficiency may take various forms, such as

- Stunted or reduced growth of the entire plant, with the plant either remaining green or lacking an overall green color with either the older or younger leaves being light green to yellow in color
- Chlorosis of leaves, either interveinal or of the whole leaf itself, with symptoms either on the younger and/or older leaves or both (chlorosis is due to the loss or lack of chlorophyll production)
- Necrosis or death of a portion (margins or interveinal areas) of a leaf, or the whole leaf, usually occurring on the older leaves
- Slow or stunted growth of terminals (rosetting), the lack of terminal growth, or death of the terminal portions of the plant
- A reddish purpling of leaves, frequently more intense on the underside of older leaves due to the accumulation of anthocyanin

A summary of symptoms of essential plant nutrient element insufficiencies is given in Table 3.6.

In some instances, a plant nutrient element insufficiency may be such that no symptoms of stress will visually appear with the plant seeming to be developing normally. This condition has been called *hidden hunger*, a condition that can be uncovered by means of either a plant analysis and/or tissue test (see Chapter 17).

A *hidden hunger* occurrence frequently affects the final yield and the quality of the product produced. For grain crops, the grain yield and quality may be less than expected; for fruit crops, abnormalities such as blossom-end rot and internal abnormalities may occur, and the post-harvest characteristics of fruits and flowers will result in poor shipping quality and reduced longevity. Another example is K insufficiency in corn, a deficiency that is not evident until maturity, when plants easily lodge.

A high N level in the plant can make the plant sensitive to moisture stress and easily susceptible to insect and disease infestations. If ammonium nitrogen ($NH_4$-N) is the primary source of N, symptoms of ammonium toxicity, fruit disorders, and the decay of conductive tissues may occur.

## TABLE 3.6
## Generalized Plant Nutrient Element Deficiency and Excess Symptoms

**Major Elements**

*Nitrogen (N)* — *Deficiency symptoms:* Light green leaf and plant color; older leaves turn yellow and will eventually turn brown and die; plant growth is slow; plants will mature early and be stunted.

*Excess symptoms:* Plants will be dark green; new growth will be succulent; susceptible if subjected to disease, insect infestation, and drought stress; plants will easily lodge; blossom abortion and lack of fruit set will occur.

*Ammonium (NH$_4$)* — *Toxicity symptoms:* Plants supplied with ammonium nitrogen (NH$_4$-N) may exhibit ammonium toxicity symptoms with carbohydrate depletion and reduced plant growth; lesions may appear on plant stems, along with downward cupping of leaves; decay of the conductive tissues at the bases of the stems and wilting under moisture stress; blossom-end fruit rot will occur and Mg deficiency symptoms may also appear.

*Phosphorus (P)* — *Deficiency symptoms:* Plant growth will be slow and stunted; older leaves will have purple coloration, particularly on the undersides.

*Excess symptoms:* Excess symptoms will be visual signs of either Zn, Fe, or Mn deficiency; high plant P content may interfere with normal Ca nutrition and typical Ca deficiency symptoms may appear.

*Potassium (K)* — *Deficiency symptoms:* Edges of older leaves will appear burned, a symptom known as scorch; plants will easily lodge and be sensitive to disease infestation; fruit and seed production will be impaired and of poor quality.

*Excess symptoms:* Plant leaves will exhibit typical Mg and possibly Ca deficiency symptoms due to cation imbalance.

*Calcium (Ca)* — *Deficiency symptoms:* Growing tips of roots and leaves will turn brown and die; the edges of leaves will look ragged as the edges of emerging leaves will stick together; fruit quality will be affected and blossom-end rot will appear on fruits.

*Excess symptoms:* Plant leaves may exhibit typical Mg deficiency symptoms; in cases of great excess, K deficiency may also occur.

*Magnesium (Mg)* — *Deficiency symptoms:* Older leaves will be yellow, with interveinal chlorosis (yellowing between veins) symptoms; growth will be slow and some plants may be easily infested by disease.

*Excess symptoms:* Results in a cation imbalance with possible Ca or K deficiency symptoms appearing.

*Sulfur (S)* — *Deficiency symptoms:* Overall light green color of the entire plant; older leaves turn light green to yellow as the deficiency intensifies.

*Excess symptoms:* Premature senescence of leaves may occur.

*(continued)*

**TABLE 3.6** (*continued*)
**Generalized Plant Nutrient Element Deficiency and Excess Symptoms**

Micronutrients

| | |
|---|---|
| *Boron (B)* | *Deficiency symptoms:* Abnormal development of growing points (meristematic tissue); apical growing points eventually become stunted and die; flowers and fruits will abort; for some grain and fruit crops, yield and quality are significantly reduced; plant stems may be brittle and easily break. |
| | *Excess symptoms:* Leaf tips and margins turn brown and die. |
| *Chlorine (Cl)* | *Deficiency symptoms:* Younger leaves will be chlorotic and plants will easily wilt. |
| | *Excess symptoms:* Premature yellowing of the lower leaves with burning of leaf margins and tips; leaf abscission will occur and plants will easily wilt. |
| *Copper (Cu)* | *Deficiency symptoms:* Plant growth will be slow; plants will be stunted; young leaves will be distorted and growing points will die. |
| | *Excess symptoms:* Iron deficiency may be induced with very slow growth; roots may be stunted. |
| *Iron (Fe)* | *Deficiency symptoms:* Interveinal chlorosis on emerging and young leaves with eventual bleaching of the new growth; when severe, the entire plant may turn light green. |
| | *Excess symptoms:* Bronzing of leaves with tiny brown spots, a typical symptom on some crops. |
| *Manganese (Mn)* | *Deficiency symptoms:* Interveinal chlorosis of young leaves while the leaves and plants remain generally green; when severe, the plants will be stunted. |
| | *Excess symptoms:* Older leaves will show brown spots surrounded by chlorotic zones and circles. |
| *Molybdenum (Mo)* | *Deficiency symptoms:* Symptoms are similar to those of N deficiency; older and middle leaves become chlorotic first and, in some instances, leaf margins are rolled and growth and flower formation are restricted. |
| | *Excess symptoms:* Not known and probably not of common occurrence. |
| *Zinc (Zn)* | *Deficiency symptoms:* Upper leaves will show interveinal chlorosis with whitening of affected leaves; leaves may be small and distorted, forming rosettes. |
| | *Excess symptoms:* Iron deficiency symptoms will develop. |

Some fungus diseases are more likely to occur on plants that are marginally deficient in a particular element, an example being the occurrence of powdery mildew on leaves of greenhouse-grown cucumber when Mg is not fully sufficient. Wheat that is insufficient in chlorine (Cl) is easily susceptible to a disease called *take-all*.

Although not generally considered an essential plant nutrient element, the lack of adequate Si in rice (possibly true of other small grains also) may cause the plants to lack stem strength and easily lodge. Silicon insufficiency has been suggested as a possible link to disease infestations, the presence of Si in plant leaves providing a barrier to the invasion of fungus hyphae into leaf cellular structures. The level of Si in the plant may be related to overall plant vigor, practically for plants being grown hydroponically when that element in not included in a nutrient solution formulation (see page 199).

The occurrence of symptoms may not necessarily be the direct effect of an essential plant nutrient element insufficiency. For example, stunted and slowed plant growth and the purpling of leaves can be the result of climatic stress, cool air and/or root temperatures, lack of adequate moisture, etc. Damage due to wind, insects, disease, and applied foliar chemicals can produce visual symptoms typical of a nutrient element insufficiency. The treatise by Porter and Lawlor (1991) describes the relationships that exist between plant growth and the plant's nutritional status to its environment.

Some nutrient element deficiencies have been classified as *physiological diseases,* such as blossom-end rot. In all these cases, carefully followed diagnostic techniques must be employed, particularly the use of plant analyses and/or tissue analyses (see Chapter 17, "Plant Analysis") if the cause for visual disorders is to be correctly identified.

An essential plant nutrient element insufficiency (deficiency or excess) can make the plant sensitive to climatic stress, and/or easily subjected to insect and disease infestations.

# 4 The Plant Root

Plant roots, their function, their ability to grow under a range of soil-climatic conditions, and the extent of soil contact will significantly affect the growth and development of the whole plant (Carson, 1993). Any impairment of root function will be quickly observed by a change in the overall physical appearance of the aerial portions of the plant, wilting being the most apparent when water uptake is restricted. A question that has no specific answer is: What degree of control has the aerial portion of the plant on the extent of the rooting system? The size and distribution of roots in the rooting medium will significantly affect plant growth, with some indication that the plant grows more in response to root growth, rather than the other way around. Depending on the plant species, restricting root growth has a varying effect on the aerial portion of the plant, referred to as the "bonsai effect." For some plant species, mainly trees, there is no radial distribution of water and essential elements within the plant itself, so root damage can be seen in reduced limb growth on the side where the root damage occurs.

## 4.1 INTRODUCTION

Plant roots provide three important functions:

1. Anchoring the plant in the rooting medium
2. Means for water absorption
3. Means for essential and nonessential plant nutrient element absorption

Every plant species has a specific root architecture; there are those with

- A single tape root with few lateral roots
- A multiple of primary roots with either a few or many branching lateral roots
- Only a fibrous root system

Each architectural form varies in the ability of the roots to occupy the soil area immediately around the plant, as well as their capability to venture into the deeper portions of the soil profile.

Factors that restrict root growth, thereby impairing plant growth and reducing plant nutrient element uptake, include

- Disease and insect damage
- Root pruning due to insect damage and cultivation procedures
- Soil compaction
- Soil acidity

- Poor drainage
- Soil temperature
- Element deficiencies
- Excess salts or Na
- Low oxygen

## 4.2   ROOT FUNCTION

Plant root function and the extent of soil contact will significantly influence the growth and development of the whole plant. Any impairment of root function will be evident as a change in the overall physical appearance of the aerial portions of the plant. Surprisingly, few roots, if fully functioning, are all that is needed to supply most or all of the water and essential plant nutrient elements needed for normal plant growth.

Roots depend on translocated photosynthates from plant leaves for their energy and structural materials (carbohydrates) necessary for growth, while the aerial portion of the plant is supplied water and absorbed elements by means of root absorption and translocation. The energy for root function comes from respiration, a process that takes place in an aerobic [oxygen ($O_2$) must be present] environment. Therefore, roots will not normally venture into anaerobic (lacking $O_2$) rooting environments, even when there is no physical restriction. Essential plant nutrient element insufficiencies can occur even though the overall soil fertility level is adequate to meet the plant's requirements.

The extent of root development and physical appearance are more a factor of the rooting medium than that associated with the plant itself.

The physical and chemical properties of the rooting environment can be modified by the plant root, thus overcoming conditions that would impact plant growth and its ability to grow under stress conditions.

## 4.3   ROOT HAIRS

Root hairs are found just behind the growing root tip, and are not a feature associated with mature roots. Root hair development is influenced by the physical and chemical characteristics around the developing root, more likely to develop when the soil is infertile and when the soil humidity is high. Root hair development is enhanced by low concentrations of nitrate ($NO_3^-$) and phosphate ($HPO_4^{2-}$ or $H_2PO_4^-$) in the soil solution, and when the rooting medium is moist, but not wet. Root hairs play a major role in the ability of the plant roots to absorb water and ions from the surrounding soil solution, as their development and presence considerably increase the absorptive root surface.

## 4.4   LATERAL ROOTS

The formation of lateral roots also increases the soil–root contact surface, and in turn, enhances root ion uptake.

## 4.5   THE RHIZOSPHERE

The rhizosphere [the narrow region of a soil that is directly influenced by root secretions and associated soil microorganisms] (www.Wikipedia.org) is the thin cylinder immediately surrounding the root, serving as the interface between the root and the rooting medium. The soil that is not part of the rhizosphere is called the bulk soil. The pH and other characteristics of the rhizosphere are different from that of the rooting medium itself. Normally, the pH of the rhizosphere is more than one unit less than the rooting medium as a whole, due to the release of $H^+$ ions from the root respiration process. This acidifying property assists in the absorption of elements that are more soluble in an acidic environment, such as P and the micronutrients, Cu, Fe, Mn, and Zn. Some plant species release what are known as "siderophores," which form complex Fe for ease of root absorption.

The rhizosphere teems with microorganism activity, one type being referred to as "mycorrhizae," a family of fungi, serving as a "buffer" zone around the root. This combined bacterial and fungal biological activity feeds on the carbonaceous materials released or sloughed off as roots move through and/or expand into the rooting medium. This significant biological activity affects the availability and uptake of ions from the soil solution into the root. This is also one of the reasons why many plants can survive in less than ideal rooting media.

## 4.6   ROOT ION ABSORPTION

The physical characteristics of the root itself have an influence on ion uptake because as the root changes anatomically, the function and rate of ion uptake are affected. In general, as the distance from the root tip increases, the rate of ion uptake decreases.

It is generally believed that a carrier system exists that literally *carries* an ion across the cell membrane and against a concentration gradient, although the specific identification of such carriers has not been determined. An ion is attached to a carrier, with the combined unit transported from the root surface into the root itself. The ion is deposited inside the root with the carrier moving back across the cell membrane to repeat the process with another ion. Another concept is that there exists an ion pump system that assists in the transport of ions across the cell membrane. In order for both of these systems to work, energy is required, which is derived from root respiration. Therefore, roots must be in an aerobic atmosphere, having access to oxygen ($O_2$) in order for ion root absorption to occur.

It is believed that a portion of the ions taken into the root do so passively. In addition, there exists what is called "free space" within the outer cells of the root where free exchange of ions occurs between those in the soil solution and those in the root. This allows some ions, such as the $K^+$ cation and the $NO_3^-$ anion to enter the root, bypassing the carrier and ion pump mechanisms for ion absorption.

Ions that carry an electrical charge may be excluded from uptake as compared to uncharged molecules; for example, as the external pH increases, the uptake of B is affected as the ratio of B as boric acid ($H_3BO_3$, uncharged molecule) to the borate anion ($BO_3^{3-}$) changes. Uptake of P is also influenced by pH as the ionic form of P in the soil solution changes from $H_2PO_4^-$ to $HPO_4^{2-}$ to $PO_4^{3-}$ as the pH increases, giving rise to a change in both the size of and the charge on the anion. By contrast, pH change has no effect on the sulfate anion ($SO_4^{2-}$); therefore, its uptake remains fairly constant with changing pH.

The effect of pH, the presence of other cations and anions, and the respiration characteristics of the root play major roles in ion uptake. In general,

- pH has a greater impact on cation than anion uptake.
- There is a greater competitiveness among the cations than anions for uptake.
- There is a charge compensation associated with the differential uptake of cations.

Also, there appears to be a *feedback* system in the root that can regulate the uptake rate of ions when

- Roots are impaired or damaged by physical circumstances (compacted soils, anaerobic conditions due to soil crusting, mechanical root pruning, etc.).
- Low soil temperature is present.
- Low soil moisture impairs ion movement in the soil by either mass. flow or diffusion.
- Excessive soil water level, creates an aerobic condition.
- Adverse biological activity such as disease and nematode infestations occurs.

## 4.7   ROOT CROPS

For some plants, their roots are the harvested part [i.e., potato (white and sweet), cassava, carrot, radish, turnip, beet, turnip, sugar beet], the root serving as the storage part for the generated photosynthate. These plants grow best in soils that are easily friable as the root expands in size.

# 5 How to be a Diagnostician

**The best fertilizer is the foot print of the farmer (grower) in his field**

*—Chinese Proverb*

## 5.1 THE DIAGNOSTIC APPROACH

The Diagnostic Approach involves taking a series of specific steps followed by an evaluation. A diagnosis of a soil/crop system may be made for evaluation purposes with no abnormalities visibly present or because there are visible signs of an abnormality. The Diagnosis Approach includes an evaluation of

- Tillage practices best suited to an area
- Appropriate crop rotation
- Moisture conservation and efficient water use
- Proper chemical environment by liming or reducing salt and salinity
- Correct amount and kind of fertilizers
- Variety or hybrid best suited for an area and to specific conditions
- Proper plant spacing
- Pest monitoring
- Herbicides for weed control
- Pesticides to control insects and diseases
- Proper method and time of planting and harvesting
- Timeliness in all operations
- Careful records and economic evaluation

## 5.2 BEING A DIAGNOSTICIAN

To make a diagnosis, the diagnostician requires experience and knowledge of the soil/crop system to be evaluated as well as curiosity.

1. The qualities of a successful diagnostician include the following:
   - The ability to go prepared with open eyes
   - Knowing how to deal with bias and given erroneous information
   - The ability to relate symptoms to cause

2. The diagnostician should have knowledge of
   - Visual essential plant element insufficiency symptoms

- Soil sampling procedures
- Plant sampling procedures
- Where to find reference material related to the soil/crop system being diagnosed

3. The diagnostician should carry the following items:
   - A spade to lift plant roots for examination
   - A knife to cut into plant stems and stalks
   - A soil sample tube to collect a soil sample for laboratory analysis and to examine the soil profile for changes in color, texture, and evidence of compaction
   - A hand lens to examine plant tissue for evidence of diseases or for insect identification
   - A notebook or electronic recording device
   - A camera to take photographs for later reference
   - Suitable containers for placing soil and plant tissue samples for later examination and/or laboratory analysis in order to maintain their integrity
   - Suitable containers if insects are collected for identification in order to maintain their integrity

4. The sequence of steps should be as follows:
   - Walk through the crop canopy
   - Make an initial evaluation of what has been observed
   - Begin to ask questions to verify what has been observed
   - Begin to eliminate possible causes from what has been observed

5. Scouting a crop:
   - Looking for insects, recording their number and species
   - Observing and recording the condition of the crop
   - Using scouting to take soil and plant tissue samples

## 5.3   DIAGNOSTIC FACTORS

A complete diagnosis includes factors associated with soil and plant characteristics, presence or absence of pests, management procedures, and weather conditions:

1. Soil factors for evaluation include
   - Soil test results: pH and level of essential elements
   - Soil salinity
   - Root zone status: soil tilth and profile depth, drainage
   - Soil surface: rough or smooth

2. Plant factors for evaluation include
   - Hybrid or variety
   - Plant spacing, population, row orientation
   - Date of planting

- Stage of growth
- Visual signs of plant stress
- Visual signs of elemental insufficiencies

3. Pest factors for evaluation include
   - Weeds present: type and quantity
   - Evidence of crop herbicide damage
   - Presence of insects: type and population
   - Presence of plant disease: on the plant or its roots

4. The management factors for evaluation include
   - Lime application: date, amount, and kind
   - Fertilizer application: amount, kind, and placement
   - Tillage practices
   - Water management

5. The climatic factors for evaluation include
   - Air temperature: sequences and extremes
   - Rainfall: amounts and distribution
   - Unusual weather events: temperature extremes, rainfall events, hail, high or low solar light intensities
   - Wind: frequency and velocity, stagnant air

## 5.4   EVALUATING DIAGNOSTIC PROCEDURES

Diagnostic procedures for evaluating an established growing crop should be used when planning a cropping program, to include

- A soil test to determine if soil acidity correction is required, applying lime in the fall (at least 3 months for planting) to correct soil acidity – until the soil when needed to mix the lime into the rooting depth
- A soil test to determine elemental needs and broadcast fertilizer to correct major insufficiencies prior to final soil preparation
- Preparing the seedbed for planting
- Following recommended procedures for planting, banding fertilizer if needed to meet a specific crop requirement
- Either prior to planting, at planting, or after planting based on procedures for best control, applying chemicals, if needed, to control weeds and insect pests
- Walking the planted field on a set schedule based on crop development
- Applying chemicals, if needed, to control weeds, and/or insects, and/or diseases
- In mid season, collecting a plant tissue sample and corresponding soil sample
- Making a careful product yield determination at various sections of the field area
- Determining product quality using procedures for that crop

## 5.5   SCOUTING

Scouting is primarily the sweeping of the plant canopy for insects and collecting them for identification and numbers. Based on the species and numbers collected, pest control chemical application made in order to minimize plant damage. There are those who specialize in this endeavor and contract with farmers (growers) to periodically scout their fields and recommend when pest control treatment is needed. Some who are in this profession also have other abilities, such as collecting soil and plant tissue samples for analysis as well as making other observations as to crop condition, soil surface characteristics, and soil moisture status.

## 5.6   WEATHER CONDITIONS

In a diagnostic evaluation of a crop, prior weather conditions may be the primary cause for the current condition of the crop. Previous weather conditions can set in motion growth characteristics that will manifest over the entire season. In corn, for example, the moisture and temperature conditions during early growth will determine ear size and kernel numbers. Plants damaged by hail during early or mid-growth may look normal weeks after damaged by hail, but the reproductive processes will be significantly impaired, resulting in the probability of poor yield and product quality.

## 5.7   FACTORS AFFECTING ESSENTIAL NUTRIENT
## ELEMENT CONCENTRATIONS IN PLANTS

Factors that affect essential nutrient element concentrations in plants include the following:

- *Soil physical factors:* soil tilth, structure, compaction, soil surface conditions
- *Soil chemical factors:* organic matter content, water pH, level of essential elements
- *Crop factors:* previous crop, date of planting, hybrid or variety, stage of growth
- *Treatment factors:* applied manures and composts; fertilizer placement; time, kind, and amount
- *Weather factors:* air temperature, rainfall (amount and time), solar light conditions, wind
- *Pest factors:* weeds and insects

## 5.8   PLANT (CROP) WILTING

Plants wilt due to one or a combination of the following factors:

- Low water availability due to lack of rain and/or applied irrigation water
- Inadequate root size and function due to poor root growing conditions in impervious soil and low aeration, poor development of a fine root structure including root hair formation
- Shallow depth of the surface soil as well as an impervious subsoil

- Hardpans and/or plow pans present within the normal root zone limiting root growth
- Root disease and presence of insects that are interfering with normal root development and function
- Soil temperature, both high and low, which reduces normal physiological functioning of the root, impairing absorption of water by plant roots
- Salinity, which reduces the absorption of water by the roots
- Weather conditions related to temperature extremes; heavy rainfall that creates an anaerobic soil condition due to soil water saturation
- High atmospheric demand conditions due to high air temperature, low relative humidity, and windy conditions

The effects of plant wilting include

- Lowering the rate of photosynthesis
- Slowing down or impairing plant growth
- Reducing fruit set, yield, and quality

## 5.9  SUMMARY

Prepare a spreadsheet giving each procedure to be followed for soil preparation, correcting soil fertility insufficiencies, cropping plan, and establishment procedures that would include

- Tillage practices best sited to that particular soil
- Appropriate crop rotation
- Proper chemical environment by liming or reducing salt and salinity
- Correct amount and kind of fertilizers
- Variety or hybrid best suited for the climatic and soil conditions
- Correct method and time of planting
- Proper plant spacing
- Moisture conservation and efficient water use
- Herbicides for weed control
- Pesticides to control insects and diseases
- Timeliness in all operations
- Pest monitoring (scouting)
- Crop monitoring (plant analysis)
- Correct method and time of harvesting

## 5.10  CERTIFIED CROP ADVISOR PROGRAMS

The Certified Crop Advisor (CCA) Program is a national voluntary certification program sponsored by the American Society of Agronomy (ASA) and the American Society for Horticultural Science (ASHS) that provides individuals with academic training, acquired skills, and demonstrated ability to give crop management advice to farmers/growers and agribusiness.

# Section II

Physical and Physiochemical
Characteristics of Soil

# 6 Soil Taxonomy, Horizontal Characteristics, and Clay Minerals

Soils have been classified in a system of soil taxonomy by soil orders and horizontal characteristics. These classifications provide useful information that relates to soil use and fertility characteristics and productivity. A more detailed discussion of this topic can be found in the book by Sumner (1980).

## 6.1 SOIL ORDERS (U.S. SYSTEM OF SOIL TAXONOMY)

**Alfisols:** Mineral soils have umbric or ochric epipedons or argillic horizons, and hold water at <1.5 MPa tension for at least 90 days when the soil is warm enough for plants to grow outdoors. Alfisols have a mean annual soil temperature of <8°C or a base saturation in the lower part of the argillic horizon of 35% or more when measured at pH 8.2. Characteristics: area 12,621 $km^2 \times 10^2$, proportion 9.6%, CEC ($mol_c$/kg) 0.12, high clay content, prominent argillic horizon, ochric horizon.

**Aridsols:** Mineral soils that have an acidic moisture regime; an ochric epipedon, and either pedogenic horizons, but no oxic horizon. Characteristics: area 912 $km^2 \times 10^2$, proportion 0.7%, CEC ($mol_c$/kg) 0.16, arid climate, prominent calcic horizon.

**Entisols:** Mineral soils that have no distinct subsurface diagnostic horizons within 1 meter of the soil surface. Characteristics: area 15,728 $km^2 \times 10^2$, proportion 12.0%, CEC ($mol_c$/kg) 0.43, lack of discernible horizons.

**Histosols:** Organic soils that have organic soil materials in more than half of the upper 80 cm, or are of any thickness if overlying rock or fragmental materials have interstices filled with organic soil minerals. Characteristics: area 1,527 $km^2 \times 10^2$, proportion 1.2%, CEC ($mol_c$/kg) 1.40, high organic content, prominent histic horizon (peat).

**Inceptisols:** Mineral soils that have one or more pedogenic horizons in which mineral materials other than carbonates, or amorphous silica have been altered or removed but not accumulated to a significant degree. Under certain conditions, inceptisols may have an ochric, histic, plaggen, or mollic epipedon. Water is available to plants more than half of the year or more than 90 consecutive days during a warm season. Characteristics: area 12,850 $km^2 \times 10^2$, proportion 9.8%, CEC (mol$_c$/kg) 0.19, young, horizons present but poorly developed.

**Mollisols:** Mineral soils that have a mollic epipedon overlying mineral material with a base saturation of 50% or more when measured at pH 7. Mollisols may have an argillic, nitric, albic, cambic, gypsic, calcic, or pettrocalcic horizon, a histic epipedon or a duripan, but not an oxic or spodic horizon. Characteristics: area 9,006 km$^2$ × 10$^2$, proportion 6.9%, CEC (mol$_c$/kg) 0.22, soft, rich, prominent mollic horizon.

**Oxisols:** Mineral soils that have an oxic horizon within 2 cm of the surface or plinthite as a continuous phase with 30 cm of the surface, and do not have a spodic or argillic horizon above the oxic horizon. Characteristics: area 9,820 km$^2$ × 10$^2$, proportion 7.5%, CEC (mol$_c$/kg) 0.05, high oxide content, prominent histic horizon, forms laterite.

**Spodosols:** Mineral soils that have a specific horizon or a plastic horizon that overlies a fragipan. Characteristics: area 3,354 km$^2$ × 10$^2$, proportion 2.5%, CEC (mol$_c$/kg) 0.11, high ash content, prominent spodic horizon.

**Ultisols:** Mineral soils that have an argillic horizon with a base saturation of <35% when measured at pH 8.2. Ultisols have a mean annual soil temperature of 8°C or higher. Characteristics: area 11,054 km$^2$ × 10$^2$, proportion 8.3%, CEC (mol$_c$/kg) 0.06, heavily leached, all horizons heavily weathered, argillic horizon present.

**Vertisols:** Mineral soils that have 30% or more clay, deep wide cracks when dry, and either gilgai micorelief intersecting slickness, or wedge-shaped structural aggregates tilted at an angle from the horizon. Characteristics: area 3,160 km$^2$ × 10$^2$, proportion 2.4%, CEC (mol$_c$/kg) 0.37, inverted, cracks, all horizons have high clay content.

## 6.2 DESIGNATIONS FOR SOIL HORIZONS AND LAYERS

**O Horizon:** Layers designated by organic material, except limnic layers that are organic.

**A Horizon:** Mineral horizons formed at the surface or below an O horizon and (1) are characterized by an accumulation of humidified organic matter intimately mixed with the mineral fraction and not dominated by properties characteristic of E or B horizons; or (2) have properties resulting from cultivation, pasturing, or similar disturbances.

**E Horizon:** Mineral horizons in which the main feature is loss of silicate clay, iron, aluminum, or some combination of these, leaving a concentration of sand and silt particles of quartz or other resistant materials.

**B Horizon:** Horizons formed below an A, E, or O horizon and are dominated by (1) carbonates, gypsum, or silica, alone or in combination; (2) evidence of removal or carbonates; (3) concentrations of sesquioxides; (4) alterations that form silicate clay; (5) formation of granular, blocky, or prismatic structure; or (6) a combination of these factors.

**C Horizon:** Horizons or layers, excluding hard bedrock, that are little affected by pedogenic processes and lack properties of O, A, E, or B horizons. Most are mineral layers, but limnic layers, whether organic or inorganic, are included.

**R Horizon:** Hard bedrock including granite, basalt, quartzite, and indurated limestone or sandstone that is sufficiently coherent to make hand digging impractical.

# 7 Physical Properties of Soils

The physical properties of a soil are determined by its textural class, structure, water-holding capacity, and drainage characteristics.

## 7.1 TEXTURAL CLASSIFICATION

Soil texture is classified into twelve classes based on the percentage quantity distribution of sand, silt, and clay (Figure 7.1). The twelve soil textural classes are

| | | |
|---|---|---|
| Clay | Sandy clay loam | Silt loam |
| Sandy clay | Silty clay loam | Silt |
| Silty clay | Loam | Loamy sand |
| Clay loam | Sandy loam | Sand |

Under some textural classification systems, the sand fraction is further divided into five classes: very coarse sand, coarse sand, medium sand, fine sand, and very fine sand.

Six of the twelve soil textural classes have been characterized based on their clay percentage content:

| Soil Texture | Approximate Clay Content (%) |
|---|---|
| Loamy sand | 5 |
| Sandy loam | 10 |
| Silt loam | 20 |
| Silty clay loam | 30 |
| Clay loam | 35 |
| Clay | >40 |

## 7.2 SOIL SEPARATES OR PRIMARY SOIL SEPARATES

The soil separates are strictly defined based on their physical size. The US System is:

| Separate | Physical Diameter (mm) |
|---|---|
| Sand | 2.0–0.05 |
| Very coarse sand | 2.0–1.0 |

| Coarse sand | 1.0–0.5 |
| Medium sand | 0.5–0.25 |
| Fine sand | 0.25–0.10 |
| Very fine sand | 0.10–0.05 |
| Silt | 0.05–0.002 |
| Clay | 0.002 |

The U.S. Department of Agriculture (USDA) classification system defines the soil separates using a different separate designation and physical diameter size:

| Separate | Physical Diameter (mm) |
| --- | --- |
| Coarse gravel | 2.0–1.0 |
| Fine gravel | 1.0–0.5 |
| Medium sand | 0.5–0.25 |
| Fine sand | 0.25–0.10 |
| Very fine sand | 0.10–0.05 |
| Silt | 0.05–0.002 |
| Clay | <0.002 |

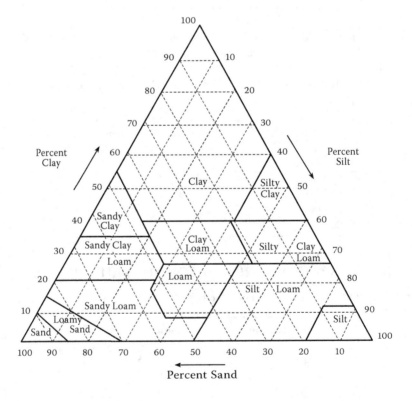

**FIGURE 7.1**    Soil textural classes based on percentage of sand, silt, and clay.

There is also an international system for size limits of soil separates:

| Fraction | Separate | Diameter Range (mm) |
|----------|----------|---------------------|
| I | Coarse sand | 2.0–0.2 |
| II | Fine sand | 0.2–0.02 |
| III | Silt | 0.02–0.002 |
| IV | Clay | <0.002 |

## 7.3   SOIL SEPARATE PROPERTIES

Soils can be divided into three categories based on their content of sand, silt, or clay:

- *Sand:* soils considered sandy are those with sand content of 50% or greater.
- *Silt:* soils that have a silt classification contain 40% or greater silt.
- *Clay:* soils that have a clay classification contain 55% or greater clay

## 7.4   SOIL TEXTURE CHARACTERIZATION DEFINITIONS

Soils with varying textural contents have been identified as

- *Fine-textured soil:* refers to those soils in which the majority of soil separates are clay and silt.
- *Coarse-textured soil:* refers to a soil in which the majority soil separate is sand.
- *Sandy soil:* inherently droughty, easy to till, less likely to compact unless containing considerable fine sand.
- *Soils high in silt:* can be compacted readily, and the soil surface can easily seal over, restricting the intake of water (rain or irrigation water easily runs off) and exchange of air.

## 7.5   SOIL STRUCTURE

Soil structure is the result of how the soil separates arrange themselves to form stable, large soil particles that determine unique physical properties, such as tilth, ease of manipulation, water infiltration and drainage rates, air exchange, etc. Particle arrangement and soil structural stability are enhanced by the content of soil organic matter that results in particle formation aggregated for ease of water infiltration and air exchange. An established soil structure can be altered by soil manipulation and compaction.

There has developed a terminology associated with particular soil structural types—such as blocky, granular, prismatic, blocky and granular—that has aided the identification of soil structural systems. A system for identifying soil structure has been developed using 5 soil designations that are associated with particular soil characteristics:

| Soil Designation | Soil Characteristics |
|---|---|
| Loose | Easy to work; neither sticky nor malleable |
| Sticky | Adheres to hands and implements |
| Plastic | Smooth and malleable, subject to smearing when cultivated |
| Friable | Breaks into crumbs under gentle pressure; not sticky; does not smear |
| Firm | Not friable even under high stress; clear; cloddy when cultivated |

The following are types of soil structure and associated soil characteristics:

- *Single grain:* loose incoherent mass of individual particles as in sands.
- *Amorphous:* coherent mass showing no evidence of any distinct arrangement of soil particles.
- *Angular blocky:* faces of peds rectangular and flattened vertices sharply angular.
- *Subangular blocky:* faces of peds subangular, vertices mostly oblique or subrounded.
- *Granular:* spheroidal peds characterized by rounded vertices.
- *Platy:* horizontal planes more or less developed.
- *Prismatic:* vertical faces of peds are well defined and edges sharp.
- *Columnar:* vertical edges near top of column are not sharp (column may be flat or round-topped or irregular).

## 7.6  TILLAGE PRACTICES

Current and long-term tillage practices can significantly affect soil tilth, either allowing for root development throughout the soil profile or restricting root development within the soil profile. Tillage procedures can assist in curbing water erosion, conserving soil water, and establishing a uniform rooting depth of soil. Tilling soils when either wet or dry, months or days before planting, will determine how easily a seeded crop will initially grow. Tillage practice, frequency and timing, can influence the rate of residue decomposition, and affect organic matter stability and soil temperature. Long-term tillage practices at a consistent depth can create what is known as a "tillage pan," a created compact soil zone that impedes water movement up or down through the soil profile as well as preventing root penetration into the soil below the pan. Pan formation can be avoided by varying the tillage depth and using different tillage implements with each tillage operation.

## 7.7  WATER-HOLDING CAPACITY

Soil texture and structure influence the amounts of water and air that are easily exchanged or move within the soil profile. These topics are discussed in detail in Chapter 22, "Soil Water, Irrigation, and Water Quality."

# 8 Physiochemical Properties of Soil

*Physiochemical* is a term that describes the combined effect of both the physical and chemical properties of a soil that define its characteristics. Many of the physiochemcial soil properties are partially defined by the percentage content of sand, silt, and clay, mineral particles of different physical size (see pages 49–50), the content and chemical properties of organic matter (see Chapter 10, "Soil Organic Matter"), and the biological activity of organisms (bacteria, fungi, worms, etc.) in the soil.

## 8.1 SOIL SEPARATE PROPERTIES

The relative percentages among the three separates—sand, silt, and clay—define the physical properties of a soil (see Chapter 6, "Soil Taxonomy and Horizontal Characteristics"), while clay, being colloidal and carrying a negative charge, giving the soil its cation exchange capacity (CEC), the amount of the charge is determined by the characteristics of the mineral colloids and organic matter content. The CEC is either expressed as milliequivalents per 100 grams (meq/100 g), or in International Units as centimoles per kilogram (cmol/kg), both being equivalent terms.

## 8.2 MAJOR PHYLLOSILICATE MINERALS IN SOILS

The CECs of the clay minerals in a soil, their structure and chemical composition based on their layer type and charge per unit formula, are shown in the following table:

| Layer Type | Group Name | Charge per Unit Formula | Common Minerals |
|---|---|---|---|
| 1:1 | Kaolinite serpentine | »0 | Kaolinite, halloysite, chyrsotile, lizardite, antigorite |
| 2:1 | Pyrophyllite talc | »0 | Pyrophyillite and talc |
| | Smectite, or montmorillinite-saponite | 0.25–0.6 | Nonmorillonite, beidellite, nontronite, saponite, hectorite, sauconite |
| | Mica | <1 | Muscovite, paragonite, biotite, phlogopite |
| | Brittle mica | »2 | Margarite, clintonite |
| | Illite | 2 | Illite |
| | Vermiculite | 0.6–1.9 | Illite |
| 2:1:1 | Chlorite | Variable | Chlorite |
| Chain | Palygorakite | — | Palygorskite, sepiotite, sepiolite |

The CEC of the commonly found colloids in soil are:

| Cation Exchange Capacity (CEC) | | |
|---|---|---|
| Type of Colloids | Mean | Range |
| | | (meq/100 g) |
| Humus | 200 | 100–300 |
| Vermiculite | 150 | 100–200 |
| Allophane | 100 | 50–200 |
| Montmorillonite | 80 | 60–100 |
| Illite | 30 | 20–40 |
| Chlorite | 20 | 10–30 |
| Kaolinite | 8 | 3–15 |
| Sesquioxides | — | 2–4 |

The relative percentage of these various colloids will determine the CEC of the whole soil. The other contributing factor to the CEC of a soil is its organic matter content, that organic matter portion which has undergone decomposition and exists in the soil as a stable substance referred to as "humus" (see page 66).

## 8.3  CATION EXCHANGE CAPACITY (CEC) OF A SOIL BASED ON TEXTURE

The CEC of a soil can be partially defined based on its textural class:

| Approximate Cation Exchange Capacity CEC) Related to Textural Classes of Soils with Water pH Levels below 7.0 | |
|---|---|
| Soil Textural Class | CEC (meq/100 g) |
| Sand | 1–8 |
| Loamy sand | 9–12 |
| Sandy or silty loam | 13–20 |
| Loam | 21–28 |
| Clay loam | 29–40 |
| Clay | >40 |

Various soil properties can be determined by the CEC (based on its clay and organic matter contents) range:

1. Soils with CEC within 11–50 meq/100 g range:
   - High clay content
   - High organic matter content
   - More aglime required to change the soil pH
   - High capacity to hold plant nutrient elements within the soil profile
   - Physical ramifications associated with a high clay content
   - High water-holding capacity

2. Soils with CEC within 1–10 meq/100 g range
- High sand content
- Low organic matter content
- Less aglime required to change the soil pH
- Low capacity of hold plant nutrient elements, loss by leaching from the soil profile
- Physical ramifications associated with high sand content
- Low water-holding capacity

## 8.4  CATION EXCHANGE CAPACITY (CEC) DETERMINATION OF A SOIL

The cation exchange capacity (CEC) of a soil can either be determined by a specific analytical procedure (Jones, 2001), or it can be estimated by knowing the content of exchangeable cations (Ca, Mg, and K) plus an estimate of the hydrogen ion ($H^+$) content based on the lime requirement of the soil. Most soil test reports give an estimation of a soil's CEC based on the cation determinations (see Chapter 16, "Soil Testing").

## 8.5  ANION EXCHANGE CAPACITY

There is also an anion exchange soil property that is associated with some unique soil types. However, the characteristics and influence of an anion exchange capacity on soil properties are not well understood, and therefore not factors that define the fertility status of a soil.

# 9  Soil pH: Its Determination and Interpretation

Soil pH is a factor that defines the "fertility" status of a soil, whose level determines the availability of most essential plant nutrient elements as well as influencing plant growth. For most acid soils, the usual soil water pH ranges between 5.5 and 6.5, with soils that are heavily weathered being closer to pH 5.5 and those less weathered nearer pH 6.5. Most crops will grow well within the soil water pH between 5.5 and 6.5. In this section the discussion relates to those soils that are naturally acid, and without liming become more acidic with time.

## 9.1  DEFINITIONS

pH is defined as the negative log of the hydrogen ion ($H^+$) concentration on a scale from 0 to 14, with 7.0 being the neutral point, and less than 7.0 defined as "acidic" and greater than 7.0 as "alkaline." An *acid soil* can be defined as "one whose water pH is less than 7.0, which is the neutral point in the pH scale" (Figure 9.1). The degree of soil acidity is determined by measuring the pH in a soil–water slurry. The soil water pH is a measure of the hydrogen ion ($H^+$) concentration in solution that is in equilibrium with those $H^+$ cations adsorbed onto the soil colloids. Colloidal material (clay and humus) can also act as if they are hydrogen ions ($H^+$), therefore contributing to soil acidity. The concentration of $H^+$ adsorbed onto the soil colloids is defined as the soil's buffer capacity (resistance to a change in pH).

## 9.2  CAUSES OF SOIL ACIDITY

A soil can become acidic as a result of natural processes, the effects of temperature and rainfall (soil profile leaching), and enhanced by the effects of cropping and crop removal, and by use of acid-forming fertilizers (see pages 145, 151). The rate of decline in pH is determined by the buffer capacity of the soil, its physiochemical properties associated with texture, structure, and soil mineral (see Chapter 8, Physiochemical Properties of Soil") and organic matter contents (see Chapter 10, "Soil Organic Matter").

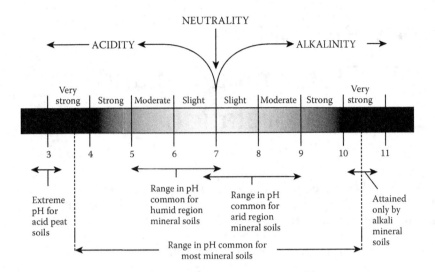

**FIGURE 9.1**   Soil pH interpretation scale for agricultural soils.

## 9.3   WATER pH DETERMINATION OF MINERAL SOIL, ORGANIC SOIL, AND ORGANIC SOILLESS ROOTING MEDIA

The determination of the water pH of a mineral soil, organic soil, or organic soilless medium is determined in a ratio-slurry with water, the ratio being specified based on the pH interpretation scale to be used. Normally for mineral and organic soils, the water-to-soil volume ratio is 1:1, while in an organic soilless mix, water is added to just the saturation point. The time between adding the water to form the slurry and making a reading is determined by the time needed to establish equilibrium, usually between 10 and 30 minutes. The pH determined in these water slurries is the result of three interacting factors:

1. Concentration of hydrogen ions ($H^+$) in solution
2. Hydrogen ions ($H^+$) in equilibrium between the colloidal faction and that in solution
3. The proton effect of the colloidal material present

If the slurry is filtered, removing the solid phase, the pH of the filtrate will be higher (can be as much as a whole pH unit) than that of the slurry, the difference in pH being determined by the physiochemical properties of the solid phase.

The pH determination for an organic soil or organic soilless rooting medium is normally determined in a slurry with water only. The optimum water pH ranges for most plants growing in three different rooting media are:

1. Mineral soil (pH 5.4–6.8)
2. Organic soil (pH 5.2–5.8)
3. Soilless organic rooting medium (pH 5.0–6.0)

Note that the optimum water pH range is not the same for these three rooting media. In addition, there is no one explanation for what effect the determined pH will have on a plant rooted in one of these three media. In general, plant roots function best in an "acidic" environment.

## 9.4   pH DETERMINATION USING A CALIBRATED PH METER

Commonly, the soil water pH is determined in a water slurry using a calibrated pH meter equipped with either separate glass and reference electrodes, or with a combination single-body electrode. Today's pH meters are rugged and easy to use. Calibration using a two-point procedure requires what are called "buffer solutions" of known pH that bracket the pH range of what is expected for the unknowns. For acid range determinations, the two buffer solutions commonly used have a pH of 4.0 and 7.0, respectively. For alkaline range determinations, pH buffers 7.0 and 10.0, respectively, are used. The procedures for calibrating a pH meter are provided with each pH meter and should be carefully followed.

Once the pH meter is calibrated, it is best to have a reference sample of known pH that is of the same matrix as that of the unknown(s). For example, when determining the pH of either a soil or soilless mix, have a soil or soilless mix available that has an already-determined pH. Some may be commercially available, or have someone using their own pH meter, prepare a reference standard. Self-generated standards based on repeated pH measurements on a bulk sample using the same pH meter as for making pH measurements on unknowns are not suitable "reference" standards.

The temperature at which the pH determination is made can affect the measured value, although measurements made at what is defined as "room temperature" will not influence electrode performance. When a pH measurement is made outside the "room temperature" range, compensation must be made for the possible temperature effect on the pH meter reading. Some pH meters are equipped with a temperature probe that corrects the pH meter determination due to varying temperature. The one caution to observe is that the temperature of the calibrating buffers be that of the unknowns.

When the electrodes are placed into the soil-water slurry, the slurry must be stirred, either using a mechanical stirrer or by moving the electrodes and/or slurry container. The standing water pH measurement can be as much as 0.5 higher than that in the moving slurry.

## 9.5   ANOTHER SOIL pH DETERMINATION PROCEDURE

The soil water pH can be determined with the use of dyes that change color with a change in pH. An aliquot of soil is placed in a depression on a white sample plate and a few drops of dye are added to the soil to saturation. After stirring, the soil is allowed to settle and the color of the ring of dye around the soil is compared to an appropriate color chart for the dye used.

## 9.6  SALT pH DETERMINATION FOR A MINERAL SOIL

For soils with low colloidal content and/or when the ion content in solution is low, pH meter readings are difficult to make because the meter does not quickly settle on a fixed point. To overcome this difficulty, the pH determination is made in a salt solution, a procedure that is coming into wider use. A salt pH determination is made in a slurry of mineral soil and salt solution, either one-hundredth molar calcium chloride (0.01M $CaCl_2$) or one normal potassium chloride (1N KCl). In such a matrix, electrical conductivity is ensured and the pH meter reading quickly stabilizes and remains fixed even when the slurry is allowed to stand. The salt pH value obtained depends on the physiochemical properties of the soil, its colloidal content, and which salt solution is used. There will be as much as a half a pH unit less than the water pH when 0.01M $CaCl_2$ is the salt solution, and as much as a whole pH unit less when determined in 1M KCl. The requirement to stir the soil–salt solution slurry is not as critical to the pH determination as when the pH is measured in a water–soil slurry. A salt pH determination is made for mineral soils only.

## 9.7  pH INTERPRETATION: MINERAL SOIL

The optimum mineral soil water pH range for best plant growth is between 5.5 and 6.5, although there are exceptions for certain soil types and plant species. Below pH 5.5, the availability of the essential elements P and Mg decline, while the concentration of Al, Mn, and for some soils Cu, begin to advance into the toxic range (see Figure 9.2). The interaction between elements, such as P and Al, result in the formation of complexes that reduce P availability. As the soil water pH increases above 6.5, the availability of Cu, Mn, Fe, and Zn declines (see Figure 9.2), and the interaction among the major elements, Ca, Mg, and K, begins to impact root absorption of both Mg and K. In most soils, the availability of P declines with increasing pH into the alkaline range (see Figure 9.2).

## 9.8  pH INTERPRETATION: ORGANIC SOILS

The optimum organic soil water pH range for best plant growth is between 5.4 and 6.0, although there are exceptions for some organic soils (depending on the physiochemical properties of the mineral and organic components) and plant species. The availability for root absorption of many of the essential plant nutrient elements declines sharply as the water pH increases above 6.4 (see Figure 9.3).

## 9.9  pH INTERPRETATION: ORGANIC SOILLESS MEDIUM

The optimum organic soilless medium water pH range for best plant growth is between 5.2 and 5.8, although there are exceptions for certain soilless media, depending on the content and properties of the major ingredients, whether there is sphagnum peat moss, pinebark, perlite, vermiculite, or compost in the mix, and the

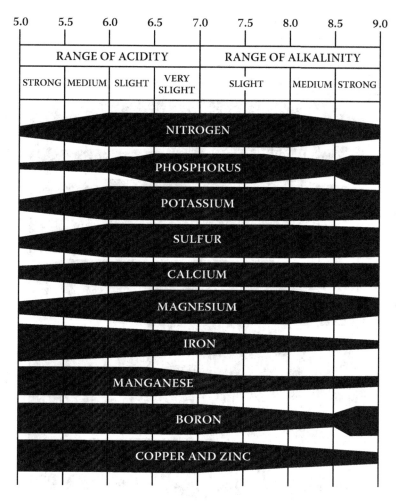

**FIGURE 9.2**   Availability of essential plant nutrient elements at different pH ranges for soils.

plant species. The availability for root absorption of many of the essential elements declines sharply as the water pH increases above 6.0.

## 9.10   SOIL pH CONSTANCY

A common procedure is to lime a soil when the soil water pH reaches a particular level of acidity, then applying sufficient lime to bring the soil to the desired pH. For most acid soils and cropping systems, the so-called ideal soil water pH is around 6.4.

Following a periodic liming schedule that results in a cycling of the soil pH can contribute to poor crop performance. Also, allowing the soil pH to cycle the subsoil will tend to be more acidic. Maintaining the soil pH of the surface soil at

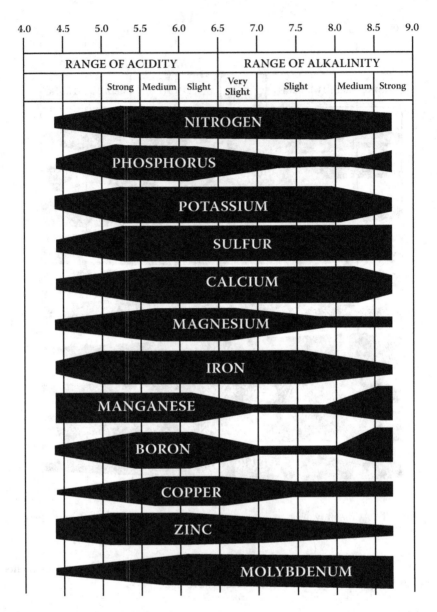

**FIGURE 9.3** Availability of essential plant nutrient elements at different pH ranges for organic soils.

the desired level by yearly liming will tend to maintain the subsoil pH, or in some instances, the soil pH of the subsoil may begin to move toward that of the surface soil. Therefore, the timing and rate, of liming should be planned to maintain a constant soil pH.

Plotting yearly soil pH determinations against time reveals the degree of cycling between lime applications. With this information, a liming program devised to minimize this cycling can be devised and, when implemented, will markedly improve crop performance.

Applying a liming material to an acid soil will not immediately bring the soil pH to its desired level, or the pH goal may never be achieved due to the intensity of natural acidifying processes when combined with that that due to cropping and crop removal, and the acidifying effects of fertilizers. By soil pH monitoring, a lime rate and schedule can be devised that will reduce the cycling of the soil pH. Frequent small-dose-rate lime applications can be effective in maintaining the soil pH once the desired level is achieved.

## 9.11   PLANT ROOT FUNCTION

The plant root has a means of buffering itself from the surrounding rooting medium; that buffer zone is known as the rhizosphere, a thin cylindrical zone encompassing the root that teems with physiochemical activity (see Chapter 4). The pH of this zone is less than that of the surrounding rooting medium, contains an active microbiological population known as microrhizophae feed by substances released from the root, and is an area where elemental ions interact, to be either absorbed by the root or removed from solution by either precipitation or chelation, and therefore made unavailable for root absorption. This latter process provides protection for the root, avoiding possible chemical injury, and by keeping substances from entering the plant root that could be detrimental to functions within the plant.

## 9.12   SOIL ACIDITY AND NPK FERTILIZER EFFICIENCY

The water pH of a soil will affect the efficiency of the fertilizer elements N, P, and K as indicated in the following table:

| Soil Acidity | Soil pH | Efficiency Percentage (%) | | |
|---|---|---|---|---|
| | | Nitrogen (N) | Phosphorus (P) | Potassium (K) |
| Extreme | 4.5 | 30 | 23 | 33 |
| Very strong | 5.0 | 53 | 34 | 52 |
| Strong | 5.5 | 77 | 46 | 77 |
| Medium | 6.0 | 89 | 52 | 100 |
| Neutral | 7.0 | 100 | 100 | 100 |

The fertilizer efficiency for P is more affected soil water pH than for fertilizer K and N.

## 9.13   SOIL pH EFFECT ON ELEMENTAL AVAILABILITY AND/OR SOIL SOLUTION COMPOSITION

With a change in pH, the concentration of some elements in the soil solution either increases or decreases, resulting in either an insufficiency if it is an essential plant nutrient element, or interference with normal root function, such as the element Al.

The effect of decreasing or increasing pH on the presence of elements in the soil solution is given as follows:

| Element | pH Decreasing | pH Increasing |
|---|---|---|
| Aluminum (Al) | Increases | Decreases |
| Copper (Cu) | Increases | Decreases |
| Iron (Fe) | Increases | Decreases |
| Magnesium (Mg) | Decreases | Increases |
| Manganese (Mn) | Increases | Decreases |
| Zinc (Zn) | Increases | Decreases |

## 9.14    SOIL BUFFER pH

The soil buffer pH is a measure of the hydrogen ion ($H^+$) concentration that exists on the soil colloids. This determination is important sense it is a measurement required for determining the lime requirement of an acid soil (see Chapter 18, "Lime and Liming Materials"). Normally, a buffer pH determination is not made if the soil water pH is greater than 6.8. There are several methods for determining the buffer pH, the method selected being determined by the cation exchange capacity (CEC) of the soil, or selected on the basis of the soil's texture. The lime requirement for a soil is a calculated value based on the use of both the water and buffer pH of an acid soil, or it can be an estimated value based on the soil water pH and soil textural class (see Chapter 18, "Liming and Liming Materials").

## 9.15    pH DETERMINATION OF WATER

The pH of ion-free water, or water containing ions that will interact with the pH glass electrode, is not easily determined. For an electrode-equipped pH meter to work, there must be electrical conductivity between the glass and reference electrodes. Conductivity is provided either by ions in solution or suspended colloidal particles that act as an anion. Without conductivity, an accurate pH determination is difficult to make; therefore, the pH value determined may have little relevance to the actual hydrogen ion ($H^+$) content in solution. The pH of pure water varies, depending on the method and time in storage.

# 10 Soil Organic Matter

The presence and nature of organic matter in a mineral soil adds unique physical and chemical properties, making such soils apparently more fertile.

The native organic matter content of a soil is determined by soil type, average mean air temperature, soil moisture level, and plant cover. For cultivated soils, it is difficult to increase the soil organic matter content, for with cultivation and cropping, the organic matter content tends to be less than that existing before cultivating and cropping. The use of green manures can temporarily increase the organic matter content, but depending on temperature, cultivation, and soil moisture level, the organic matter content of a soil will tend to stabilize at a certain content. At what content stabilization occurs also depends on the cropping system, with low organic matter soil content stabilization occurring with row crop culture systems versus pasture or hay culture systems, or for long-term cropping systems. The range of organic matter content is between 1 and 5%, with most soils being at the 1 to 2% level.

High organic matter soil content is frequently associated with that being thought to be more fertile than a similar soil with a lower content—a judgment that is not true. It is believed by some that the organic matter in the soil is associated with plant health, a healthy soil being one containing a substantial quantity of organic matter. The need to add a source of organic matter to either increase or maintain a certain soil organic matter content is considered a desirable practice to ensure the maintenance of a healthy soil. The requirements for establishing and maintaining a productive soil are discussed in Chapter 2.

There are soils that are defined as organic in that more than 20% of the soil is in the form of some organic material. Such soils have unique properties and are used for the growing of particular crops. Organic soils represent less than 5% of all cropped soils.

## 10.1 DEFINITIONS OF SOIL ORGANIC MATTER AND ITS COMPONENTS

### 10.1.1 DEFINITIONS

**Soil organic matter:** The sum total of all natural and thermally altered biologically derived organic material found in the soil or on soil surfaces irrespective of its source, whether it is living or dead, or stage of decomposition, but excluding the above-ground portions of living plants.

**Living components:** Living tissues of plant origin. Standing plant components that are dead.

**Phytomass:** (For example, standing dead trees) are also considered phytomass.

**Microbial biomass:** Organic matter associated with the cells of living microorganisms.

**Faunal biomass:** Organic matter associated with living soil fauna.

**Nonliving components:** Organic fragments with recognizable cellular structure derived particulate organic matter from any source but usually dominated by plant-derived materials.

**Liter:** Organic materials derived from mineral residues located on the soil surface.

**Macroorganic matter:** Fragments of organic matter >20 mm or >50 mm (i.e., greater than the lower size limit of the sand fraction) contained within the mineral soil matrix and typically isolated by sieving a dispersed soil.

**Light fraction:** Organic materials isolated from mineral soils by flotation of dispersed suspensions on water or heavy liquids of densities 1.5 to 2.0 mg/m$^2$.

**Dissolved organic matter:** Water-soluble organic compounds found in the soil solutions that are less than 0.45 mm by definition. Typically, this fraction consists of simple compounds of biological origin (e.g., metabolites of microbial and plant processes), including sugar, amino acids, low-molecular-weight organic acids (e.g., citric, malic, etc.) but may also include large molecules.

**Humus:** Organic materials remaining in the soil after removal of macroorganic matter and dissolved organic matter.

**Non-humic:** Identifiable organic structures that can be placed into discrete

**Biomolecules:** Categories of biopolymers, including polysaccharides and sugars, proteins and amino acids, fats, waxes and other lipids, and lignin.

**Humic substances:** Organic materials with chemical structures that do not allow them to be placed into the category of non-humic biomolecules.

**Humic acids:** Organic materials that are soluble in alkaline solution but precipitate on acidification of the alkaline extract.

**Fulvic acid:** Organic materials that are soluble in alkaline solution and remain soluble on acidification of the alkaline extracts.

**Humin:** Organic materials that are insoluble in alkaline solution.

**Inert organic matter:** Highly carbonized organic materials, including charcoal, charred plant materials, graphite, and coal with long turnover times.

## 10.2 HUMUS

A portion of soil organic matter exists as humic compounds. These compounds are defined as amorphous, colloidal, polydispersed substances and yellow to brown-black in color. They are hydrophilic, acidic, and high in molecular weight, ranging from several hundreds to thousands of atomic units. Humus is the primary substance contributing to the cation exchange capacity (CEC) of a soil. It can also act as a chelate in the soil solution, thus affecting root ion absorption.

## 10.3 SOIL ORGANIC MATTER CHARACTERISTICS

### 10.3.1 PHYSICAL CHARACTERISTICS

- Improves soil tilth and structure by stabilizing soil aggregates depending on soil texture, temperature, rainfall, aeration, and fertility level
- Reduces soil compaction and surface crusting
- Increases water infiltration and aeration
- Increases water-holding capacity
- Reduces soil erosion

### 10.3.2 PHYSICOCHEMICAL CHARACTERISTICS

- Increases the CEC, higher per weight than inorganic colloids
- Buffers the soil against rapid changes in pH
- Source of supply for the essential elements N, P, S, and B

### 10.3.3 BIOLOGICAL CHARACTERISTICS

- Serves as an energy source for soil organisms

### 10.3.4 SOURCES OF SOIL ORGANIC MATTER

- Plant residues
- Soil organism residue

### 10.3.5 CONTENT

- Increased or decreased by cropping patterns
- Crop residue management
- Frequency of tillage
- Soil fertility practices

## 10.4 METHODS OF SOIL ORGANIC MATTER DETERMINATION

There are various laboratory methods for determining the organic content of a soil, each giving different results depending on the methodology. Loss-on-ignition (LOI) is widely used; the ignition temperature and length of ignition are critical parameters. Soil preparation is an important parameter, as a portion of the LOI measurement may be due to loss of soil-bound water. Various wet oxidation procedures are used, each method giving slightly different results depending on the reagent used and also the laboratory procedure. All these procedures are described by Jones (1998).

## 10.5  MANAGEMENT REQUIREMENTS FOR HIGH ORGANIC MATTER CONTENT SOILS

Soils high in organic matter content (>5%) are more difficult to manage because they remain wet and cool when soil preparation procedures are required. High organic matter soils tend to be sticky when wet; and when plowed when wet, clods easily form that can be difficult to break up, requiring additional tilling in preparation for seeding or planting. Remaining cool with rising air temperatures, delayed seeding or planting may be required, and then after seeding or planting, plant growth will be slow due to cool soil temperatures.

## 10.6  ADVERSE EFFECTS OF ORGANIC MATTER ADDITIONS

The frequent addition of organic materials, such as composts, animal manures, sewage sludge, organic waste products, etc., to a mineral soil, with time, have an adverse effect. With increasing organic matter content, a soil will have higher water-holding capacity, will remain cool with warming air temperatures and make the soil more difficult to cultivate when wet. Depending on the source of the organic material, elements may be added that would accumulate, and with time adversely affect the nutritional status of the soil and growing crop. For example, poultry manure contains substantial quantities of K that, with time, will adversely alter the cation ratio in the soil, eventually inducing either an Mg or Ca plant deficiency. Some organic waste products contain elements, particularly heavy metals, that can become toxic to plants or adversely affect the quality of produced product. It is advisable that an organic material before being added to a soil be assayed to determine its elemental content. Based on the findings, apply at rates and times to prevent any adverse effects from occurring (see Chapter 13, "Elements Considered Toxic to Plants").

# Section III

---

*Plant Elemental Requirements and Associated Elements*

# 11 Major Essential Plant Elements

Nine elements have been designated as major elements: C, H, O, N, P, K, Ca, Mg, and S—elements that have been established as essential to plants and that are found in percent (%) concentrations in plant dry matter. The criteria for designating an element as essential are discussed in Chapter 3. This chapter provides a discussion of major essential elements in plants.

## 11.1 TERMINOLOGY

The major elements have also been designated as *macroelements* or *macronutrients*, conforming to the designation given for micronutrients. However, the term major element is preferred to designate these nine elements.

There are other designations currently used or have been used in the past for several of the major elements. For example, N, P, and K are frequently referred to as the *fertilizer* elements because they are the primary elements in commonly formulated chemical fertilizers. In fertilizers, the P and K contents are given as their oxides, $P_2O_5$ and $K_2O$, and are listed in the order $N–P_2O_5–K_2O$ on the fertilizer bag or container. For specialty fertilizers, a fourth number may be given that would refer to either the Mg or S percent content (see Chapter 19, "Inorganic Chemical Fertilizers and Their Properties").

In the past, Ca, Mg, and S were called *secondary* elements, a term that is no longer accepted for identifying these three major elements.

Another category can be made for C, H, and O, which are the primary constituents of the carbohydrate fraction in the plant, constituting the structural compounds.

## 11.2 METHODS OF EXPRESSION

Five of the major elements, normally designated as percent concentrations in the plant's dry matter, can be expressed in SI units as milligrams per kilogram (mg/kg), centimoles ($p^+$) per kilogram [cmol ($p^+$)/kg], or centimoles per kilogram (cmol/kg). These unit comparisons are shown below for the five major elements (levels selected for illustrative purposes):

| Element | Percent | g/kg | cmol(p⁺)/kg | cmol/kg |
|---------|---------|------|-------------|---------|
| P | 0.32 | 3.2 | — | 10 |
| K | 1.95 | 19.5 | 50 | 50 |
| Ca | 2.00 | 20.0 | 25 | 50 |
| Mg | 0.48 | 4.8 | 10 | 20 |
| S | 0.32 | 3.2 | — | 10 |

## 11.3   ESTABLISHED DATE FOR ESSENTIALITY/RESEARCHERS

All of the major elements were identified as essential (required by plants to grow and complete their life cycle) before the criteria for essentiality were established by Arnon and Stout (1939)—criteria that have been and are accepted today by the research community.

## 11.4   CARBON, HYDROGEN, AND OXYGEN

These three major elements are combined in green plants in the process called *photosynthesis*:

$$\text{Carbon dioxide } (6CO_2) + \text{Water } (6H_2O)$$

$$\downarrow$$

$$(\text{in the presence of light and chlorophyll})$$

$$\downarrow$$

$$\text{Carbohydrate } (C_6H_{12}O_6) + \text{Oxygen } (O_2)$$

that is, the conversion of light energy to chemical energy. For active photosynthesis to take place, the

- Plant must be fully turgid (not under water stress.
- Stomata must be open, which enhances gaseous exchange so that $CO_2$ can readily enter plant leaves.
- Leaf surface must be exposed to full sunlight.
- Plant must be nutritionally sound.

Two major essential elements, Mg [a constituent of the chlorophyll molecule (see page 16) and P and Fe play direct roles in photosynthesis. If these elements are not present at their sufficiency levels, photosynthetic activity will be significantly reduced. For further discussion on photosynthesis, see Chapter 3.

## 11.5   MAJOR ESSENTIAL ELEMENT PROPERTIES

### 11.5.1   NITROGEN (N)

**Established date for essentiality/researchers:**
- 1802 (de Saussure) and 1851–1855 (Boussingault)

**Functions in plants:**
- Found in both inorganic and organic forms in the plant
- Combines with C, H, O, and sometimes S, to form amino acids, amino enzymes, nucleic acids, chlorophyll, alkaloids, and purine bases
- Organic N predominates as high-molecular-weight proteins in plants
- Inorganic N can accumulate in the plant, primarily in stems and conductive tissue, in the nitrate ($NO_3^-$) form

**Content and distribution:**

- Consists of 1.50% to 6.00% of the dry weight of many crop plants with sufficiency values between 2.50 and 3.50% in leaf tissue.
- A lower range of 1.80% to 2.20% is found in most fruit crops and a higher range of 4.80% to 5.50% in legume species.
- Critical values vary considerably, depending on plant species, stage of growth, and plant part.
- See pages 136–137 for a listing of critical values and sufficiency ranges for a number of crop plants.
- Highest concentrations are found in new leaves, with the total plant N content normally decreasing with the age of the plant or any one of its parts. Ammonium ($NH_4$)-fertilized plants are usually higher in combined N than those that have mostly $NO_3$-N available for adsorption and utilization.
- High-yielding crops will contain 50 to 500 lbs N/A (56 to 560 kg N/ha), with the extent of removal dependent on the disposition of the crop.
- Exists as the nitrate anion ($NO_3^-$) in main stems and leaf petioles, ranging in concentration from 8,000 to 12,000 ppm during early growth, and declines to the range from 3,000 to 8,000 ppm in mid-season; most concentrated at the base of the main stem and in the petioles of recently fully matured leaves, with its determination in either stem or petiole tissue used as a means of determining the N status of plants or as a means of regulating N supplemental fertilizer applications.
- Soluble amino acids are also found in plant tissues.

**Interaction with other elements:**

- Relationship between N and P in plants is well known, as is the relationship between N and K, the ratios of N and P and N and K being used as DRIS norms for interpreting a plant analysis result (see Beverly, 1991).
- Uptake of $NO_3$ stimulates the uptake of cations, while chloride ($Cl^-$) and hydroxyl ($OH^-$) anions restrict $NO_3$ anion uptake.
- High carbohydrate status enhances the uptake of ammonium ($NH_4$).
- Uptake of $NH_4$ restricts cations, which can lead to Ca deficiency, as well as to reduced K levels in the plant.

**Available forms for root absorption**

- Exists in the soil solution as either the $NO_3^-$ anion or the $NH_4^+$ cation, the uptake of either form influenced by soil pH, temperature, and the presence of other ions in the soil solution.
- The $NH_4^+$ cation participates in cation exchange in the soil; nitrite ($NO_2^-$) may be present in the soil solution under anaerobic conditions and is toxic to plants at very low levels (<5 ppm).

**Movement in soil and root absorption:**

- The $NO_3^-$ anion moves in the soil primarily by mass flow with most of the $NO_3^-$ absorbed when it reaches the root surface.
- Nitrate ions can be readily leached from the rooting zone by irrigation water and/or rainfall, or lifted into the rooting zone by upward moving

water, driven by water loss as a result of evaporation at the soil surface and/
or evapotranspiration by plants.

- The $NH_4^+$ cation acts much like the $K^+$ cation in the soil, and its movement in the soil solution is primarily by diffusion.

**Deficiency symptoms:**

- Plants deficient in N are very slow growing, weak, and stunted.
- Typically, plants are light green to yellow in foliage color, starting with the older or mature foliage.
- Initial and more severe symptoms of yellow-leaf deficiency are seen in older leaves, as N is mobilized in the older tissue for transport to the actively growing portions of the plant.
- N-deficient plants will mature early with yield and quality significantly reduced.

**Excess (toxicity) symptoms:**

- Plants with an excess of N are dark green in color with succulent foliage, which is easily susceptible to disease and insect invasion.
- Plants may lodge easily, are susceptible to drought stress, and fruit and seed crops may fail to yield.
- Produced fruit and grain will be of poor quality.
- If $NH_4$ is the only or major form of N available for plant uptake, a toxicity condition may develop that results in a breakdown of vascular tissue, thereby restricting water uptake.
- Fruiting crops (e.g., tomato, pepper, cucumber) may develop blossom-end rot symptoms, or fruit set may be poor.
- Symptoms of Ca deficiency may occur if $NH_4$ is the primary source of N.
- Carbohydrate depletion can occur with $NH_4$ nutrition, resulting in growth reduction.

**Fertilizer sources:**

- A list of N-containing chemical fertilizers is given in Table 19.1.
- Response to applied chemical N fertilizer would be expected to be similar, irrespective of the N source, if applied as directed.
- The efficiency of N utilization will vary with time, method of application, and N form.
- Side effects of N fertilizer application is the acidifying effect of ammonium-form fertilizers (see page 151).
- Accumulation of N in the soil profile from the addition of chemical N fertilizers can contribute to the accumulation of N in ground and surface waters.

### 11.5.2 Phosphorus (P)

**Established date for essentiality/researchers:**

- 1839 (Liebig) and 1861 (Ville)

**Functions in plants:**

- A component of certain enzymes and proteins, adenosine triphosphate (ATP), ribonucleic acids (RNA), deoxyribonucleic acids (DNA), and phytin.

- ATP is involved in various energy transfer reactions, and RNA and DNA are components of genetic information.

**Content and distribution in plants:**

- Consists of 0.15% to 1.00% of the dry weight of most plants with sufficiency values from 0.20% to 0.40% in recently mature leaf tissue.
- Critical values for P are normally less than 0.20% (when deficient) and greater than 1.00% (when in excess).
- See pages 136–137 for a listing of critical values and sufficiency ranges for a number of crop plants.
- Content in leaves tends to decrease with age.
- Highest concentration found in new leaves and their petioles.
- High-yielding crops contain from 15 to 75 lbs P/A (17 to 84 kg P/ha).
- Amount of P present when crops are harvested will be considerably less for grain crops when only the grain is removed, leaving behind most of the P in the remainder of the plant.
- Soluble P (in 2% acetic acid) present as the orthophosphate ($PO_4^{3-}$) anion in main stems and leaf petioles of the actively growing portions of the plant, ranges from 100 to 5,000 ppm of the dry weight and can be used to evaluate the P status of the plant; critical concentrations occur at approximately 2,500 ppm.

**Interaction with other elements:**

- Elemental ratio of 3-to-1 between N and P is considered critical.
- Relationships exist between P and Cu, Fe, Mn, and zinc (Zn), with a 200-to-1 ratio between P and Zn being critical.
- Ratio of N to P is used as a DRIS norm for interpreting a plant analysis (see Beverly, 1991).

**Available forms for root absorption:**

- Exists in most soils in about equal amounts of organic or inorganic forms.
- Dihydrogen phosphate ($H_2PO_4^-$) and monohydrogen phosphate ($HPO_4^{2-}$) are the two anion forms of P in the soil solution, which form and their concentration depending on soil water pH.
- Al, Fe, and Ca phosphates are the major inorganic sources of P, the relative amount among these three forms being a function of soil water pH.
- Release of P into the soil solution with the decomposition of crop residues and microorganisms can be a major source of P for plant utilization.
- Plant availability is influenced by soil water pH (see Figures 9.1 and 9.2).

**Movement in soil and root absorption:**

- Phosphate ($H_2PO_4^-$ and $HPO_4^{2-}$) anions are brought in contact with the root surface primarily by diffusion in the soil solution.
- Root interception and the abundance of root hairs will significantly increase the opportunity for P absorption.
- Cool soil temperatures and low soil moisture contents can reduce P uptake, and therefore create a P deficiency.

**Deficiency symptoms:**
- Slow-growing, weak, and stunted plants that may be dark green in color with older leaves showing a purple pigmentation are symptomatic of P deficiency.
- Being fairly mobile in the plant, P-deficiency symptoms initially occur in the older tissue.

**Excess (toxicity) symptoms:**
- An excess of P appears mainly in the form of a micronutrient deficiency, with either Fe or Zn being the first elements to be affected.
- High P content may also interfere with the normal metabolism of the plant.
- Leaf contents of P greater than 1.00% are generally considered toxic.
- List of P-containing chemical fertilizers is given in Table 19.1.
- P-containing fertilizers vary considerably in their water solubility, which can affect crop response.
- Method of application (broadcast versus row) will also influence availability as applied P can be readily fixed into unavailable forms by elements (Al) and other substances in the soil.
- P accumulation in cropland soils above that needed by most crop plants is becoming a potential source of water pollution when either leached or carried in eroded soil materials into ground and surface waters.

## 11.5.3   POTASSIUM (K)

**Established date for essentiality/researchers:**
- 1866 (Bimer and Lacanus)

**Functions in plants:**
- Involved in maintaining the water status of the plant, the turgor pressure of its cells, and the opening and closing of its stomata.
- Required for the accumulation and translocation of newly formed carbohydrates.

**Content and distribution in plants**
- Consists of 1.00% to 5.00% of the dry weight of leaf tissue with sufficiency values from 1.50% to 3.00% in recently mature leaf tissue for many crop plants.
- Considered deficient or in excess when K critical values are less than 1.50% or greater than 5.00%, respectively.
- See pages 136–137 for a listing of critical values and sufficiency ranges for a number of crop plants.
- When in excess, K levels may exceed the sufficiency level by two- to three-fold.
- Sufficient K can be as high as 6.00% to 8.00% in the stem tissue of some vegetable crops.
- Highest concentrations are found in new leaves, their petioles, and plant stems, content in leaves decreasing with age.
- High-yielding crops contain from 50 to 500 lbs K/A (56 to 560 kg/ha), with crops, such as banana containing 1,500 lbs/A (1,680 kg/ha).

- Most plants will absorb more K than they need; this excess is frequently referred to as *luxury consumption.*
- Harvest of most fruits removes sizable quantities of K from the soil.
- Because K does not exist in combined form in the plant, it can be extracted easily from fresh or dried tissue, the extracted concentration essentially equals that of the total, with some vegetable crops considered K deficient when extracted sap from fresh stems and petioles contain less than 2,000 ppm K, and adequate when the K content is greater than 3,000 ppm.

**Interaction with other elements:**
- Relationship between K and Mg is well known, as is the relationship between K and Ca; high K plant contents first result in a Mg deficiency, and when K is in greater imbalance, will cause a Ca deficiency.
- The K-to-Mg and K-to-Ca ratios are used as DRIS norms for interpreting a plant analysis result (see Beverly, 1998).
- The ammonium ($NH_4^+$) cation plays a role in the balance that exists among the three cations, $K^+$, $Ca^{2+}$, and $Mg^{2+}$.

**Available forms for root absorption:**
- As the $K^+$ cation in the soil solution.
- As exchangeable $K^+$ adsorbed to soil colloids.
- As fixed K in the lattice of 2:1 clays.
- As a component in K-bearing minerals.
- An equilibrium exists between K in the soil solution, exchangeable K, and fixed K.
- When K fertilizer is applied to the soil, the equilibrium shifts toward exchangeable and fixed K, a shift that is reversed as K is removed from the soil solution by root absorption.
- Plant availability is influenced by soil water pH (see Figures 9.2 and 9.3).

**Movement in soil and root absorption:**
- Moves to the root-absorbing surface by diffusion in the soil solution, the rate of diffusion highly temperature dependent.
- The extent of root contact (root density) with the soil also has a significant effect on uptake.
- Soil oxygen ($O_2$) has a greater effect on K uptake than for most of the other ions in the soil solution.

**Deficiency symptoms:**
- Plants will lodge easily.
- Are sensitive to disease infestation.
- Fruit yield and quality will be reduced.
- Older leaves will look as if they had been burned along the edges, a deficiency symptom known as *scorch*, as K is mobile in the plant.
- Deficiency symptoms first appear in older plant tissue, as K is mobile in the plant.
- K-deficient plants may also become sensitive to the presence of $NH_4$, leading to a possible $NH_4$ toxicity syndrome.

**Excess (toxicity) symptoms:**
- Will become deficient in Mg and possibly Ca, due to the imbalance.
- Mg deficiency is most likely to occur first.

**Fertilizer sources:**
- List of K-containing chemical fertilizers is given in Table 19.1.

## 11.5.4  CALCIUM (CA)

**Established date for essentiality/researchers:**
- 1862 (Strohmann)

**Functions in plants:**
- Plays an important part in maintaining cell integrity and membrane permeability; enhances pollen germination and growth.
- Activates a number of enzymes for cell mitosis, division, and elongation.
- May also be important for protein synthesis and carbohydrate transfer.
- Its presence may serve to detoxify the presence of heavy metals in the plant.

**Content and distribution in plants:**
- Content in plants ranges between 0.20 and 5.00% of the dry weight in leaf tissue, with sufficiency values from 0.30% to 3.00% in the leaf tissue of most crop plants.
- Critical values for Ca vary considerably among various crop species, lowest for the grain crops and highest for some vegetable and most fruit crops.
- See pages 136–137 for a listing of critical values and sufficiency ranges for a number of crop plants.
- Highest concentrations are found in older leaves as the Ca content of leaves tends to increase with age.
- High-yielding crops contain from 10 to 175 lbs Ca/A (11 to 196 kg Ca/ha).
- Ca removal will be considerably less for grain and most fruit crops when only the grain or fruit is removed, leaving behind the plant, which contains most of the Ca.
- It has been suggested that total Ca content does not relate to sufficiency, as Ca accumulates in some plants as crystals of calcium oxalate; therefore extractable Ca (in 2% acetic acid) may be a better indicator of sufficiency.
- The critical Ca concentration for soluble Ca is around 800 ppm, a concentration of Ca that has been suggested as the true *critical* value for most plants.

**Interaction with other elements:**
- Relationship between Ca and K is as well known as that between Ca and Mg; these ratios are used as DRIS norms for the interpretation of a plant analysis result (see Beverly, 1991).
- Ratio of Ca to N in fruit crops and a similar ratio between Ca and B may be related to quality.
- Ammonium nutrition can create a Ca deficiency by reducing Ca uptake.

**Available forms for root absorption:**
- Calcium exists as the $Ca^{2+}$ cation in the soil solution and as exchangeable Ca on soil colloids.

- Usually the cation of highest concentration in the soil in both soluble and exchangeable forms for soils high in pH (>8.0), soils that may contain sizable quantities of Ca as precipitates of calcium carbonate ($CaCO_3$) and calcium sulfate ($CaSO_4$).
- It is generally assumed that if the soil pH is within the acceptable range in the rooting media, Ca should be of sufficient concentration to ensure plant-Ca sufficiency, assuming that other factors are also within their normal ranges.
- Soil pH seems to have little effect on Ca uptake (see Figures 9.2 and 9.3).

**Movement in soil and root absorption:**
- Moves in the soil by mass flow, the dominant supply factor, and also by diffusion.
- Availability can be significantly affected by soil moisture level.
- Reduced plant evapotranspiration will also reduce the uptake of Ca by the plant.

**Deficiency symptoms:**
- Growing tips of roots and leaves of Ca-deficient plants turn brown and die, a symptom frequently referred to as *tip burn*.
- Leaves curl and their margins turn brown with newly emerging leaves sticking together at the margins, leaving the expanded leaves shredded on their edges.
- Fruit quality will be reduced with a high incidence of blossom-end rot and internal decay.
- Being relatively immobile in the plant, deficiencies occur at the growing terminals.
- Reproduction may be delayed or terminated altogether.
- Conductive tissue at the base of the plant will decay, resulting in a reduction of the uptake of water, wilting on high atmospheric demand days, and a reduction in essential element uptake.

**Excess (toxicity) symptoms:**
- Excessive Ca content will produce a deficiency of either Mg or K, depending on the concentration of these two elements in the plant.

**Fertilizer sources:**
- It is generally assumed that maintaining the pH of an acid soil within the optimum range (5.8 to 7.5) by frequent liming will provide sufficient Ca to meet crop requirements.
- Sources of Ca for soil application are given in Table 19.1.

### 11.5.5 MAGNESIUM (MG)

**Established date for essentiality/researchers:**
- 1875 (Boehm)

**Functions in plants:**
- A component of the chlorophyll molecule (see Figure 3.1, page 16).

- Serves as a cofactor in most enzymes that activate phosphorylation processes as a bridge between pyrophosphate structures of ATP or ADP and the enzyme molecule.
- Stabilizes the ribosome particles in the configuration for protein synthesis.

**Content and distribution:**

- Plant content ranges between 0.15 and 1.00% of the dry weight in leaf tissue, with the sufficiency value being 0.25% in the leaf tissue of most crop plants.
- Critical values will vary among various crop species, being lowest for the grain crops and highest for legumes and some vegetable and fruit crops.
- See pages 136–137 for a listing of critical values and sufficiency ranges for a number of crops.
- Content in leaves increases with age, the highest concentrations being found in older leaves.
- High-yielding crops will contain 10 to 175 lbs Mg/A (11 to 196 kg Mg/ha), with crop removal being considerably less for grain and some fruit crops when only the grain or fruit is removed, leaving behind most of the Mg that exists primarily in the plant itself.
- Some plant species or cultivars within species have a particular sensitivity to Mg, becoming Mg deficient under moisture and/or temperature stress even though Mg may be at sufficient availability levels in the rooting media.

**Interaction with other elements:**

- Relationship between Mg and K is well known, as is the relationship between Mg and Ca.
- Ratios are used as DRIS norms for the interpretation of a plant analysis result (see Beverly, 1991).
- Mg deficiency can be induced by high concentrations of either the $NH_4^+$, $K^+$, or $Ca^{2+}$ cations in the rooting medium, as the $Mg^{2+}$ cation is the poorest competitor among these three cations.

**Available forms for root absorption:**

- Exists as the $Mg^{2+}$ cation in the soil solution.
- As exchangeable Mg on soil colloids, usually the cation being next to the highest in concentration in the soil in both soluble and exchangeable forms when the soil is slightly acid to neutral in pH.
- Availability declines significantly when the soil water pH is less than 5.4 (see Figures 9.2 and 9.3).

**Movement in soil and root absorption:**

- Supply to the roots depends on root interception, mass flow, and diffusion, with mass flow being the primary delivery mechanism.
- Deficiency can occur under soil moisture stress even when the soil is adequate in available Mg.

**Deficiency symptoms:**

- A yellowing of leaves or interveinal chlorosis, which begins on the older leaves because Mg is a mobile element in the plant.

- With increased deficiency, symptoms will also appear on the younger leaves with the development of necrosis symptoms when the deficiency is very severe.

**Excess (toxicity) symptoms:**
- No specific toxicity symptoms, as the Mg plant content can be quite high (~1.0%) in leaf tissue without inducing a deficiency of either Ca or K.
- An imbalance among Ca and K with Mg occurs when the Mg content in the plant is unusually high (>1.0%), which may reduce growth due to the imbalance.

**Fertilizer sources:**
- For acid soils, and as for Ca, it is generally assumed that maintaining the soil pH within the optimum range (5.8 to 7.5) by frequent liming using dolomitic (Mg-bearing) limestone or other high-content-Mg liming materials, will provide sufficient Mg to meet crop requirements.
- Sources of Mg for soil application are given in Table 19.1.

## 11.5.6 SULFUR (S)

**Established date for essentiality/researchers:**
- 1866 (Bimer and Lucanus)

**Functions in plants:**
- Involved in protein synthesis.
- Is part of the amino acids cystine and thiamine.
- Is present in peptide glutathione, coenzyme A, and vitamin $B_1$, and in glucosides such as mustard oil and thiols, which contribute the characteristic odor and taste to plants in the *Cruciferae* and *Liliaceae* families.
- Reduces the incidence of disease in many plants.

**Content and distribution in plants:**
- Content in leaf tissue ranges from 0.15% to 0.50% of the dry weight, total S content varying with plant species and stage of growth.
- See pages 136–137 for a listing of critical ranges and sufficiency ranges for a number of crops.
- Some plants may contain from 10 to 80 lbs S/A (11 to 90 kg S/ha), with cereals, grasses, and potato removing approximately 10 lbs S/A, while sugar beet, cabbage, alfalfa, and cotton will remove from 15 to 40 lbs S/A (17 to 45 kg S/ha).

**Interaction with other elements:**
- The N-to-S ratio may be as important as total S alone or the ratio of sulfate-sulfur ($SO_4$-S) to total S as indicators of S sufficiency.
- *Cruciferae* accumulate three times as much S as P.
- *Leguminosae* accumulate equal amounts of S and P.
- Cereals accumulate one-third less S than P.

**Available forms for root absorption:**
- Over 90% of available S exists in the soil organic matter, which has an approximate 10:1 nitrogen:sulfur (N:S) ratio available, depending on organic matter decomposition rates.

- The sulfate ($SO_4^{2-}$) anion is the primary available form found in the soil solution.
- In general, most of the available $SO_4$ is found in the subsoil as the anion and can be easily leached from the surface horizon.
- Availability may depend on that deposited in rainfall (acid rain) and/or that released from organic matter decomposition.
- At high soil pH (>7.0), S may be precipitated as calcium sulfate ($CaSO_4$), while at lower pH levels (<4.0), the $SO_4^{2-}$ anion may be adsorbed by Al and Fe oxides.
- Plant availability is influenced by soil water pH (see Figures 9.2 and 9.3).

**Movement in soil and root absorption:**
- Moves in the soil as the $SO_4^{2-}$ anion by mass flow and within the soil solution by diffusion.
- Low soil-moisture conditions can inhibit S uptake.
- Sulfate may precipitate as calcium sulfate ($CaSO_4$) around the roots if mass flow brings $SO_4^{2-}$ anions at a rate greater than what can be absorbed.

**Deficiency symptoms:**
- Light yellow-green in foliage color initially over the entire plant.
- Roots are longer than normal and stems become woody; also, root nodulation in legumes is reduced and delayed maturity occurs in grains.
- Interestingly, S deficiency is desired in tobacco in order to obtain proper leaf color.
- S-deficiency symptoms can sometimes be confused with N-deficiency symptoms, although S symptoms normally affect the whole plant, while N-deficiency symptoms occur initially on the older portions of the plant.
- Frequently on sandy and/or acid soils, symptoms of S deficiency may occur in newly emerging plants only to disappear as the plant roots enter the subsoil because S as $SO_4^{2-}$ tends to accumulate in the subsoil under such soil conditions; drought conditions may reduce the uptake of S, thereby inducing a S deficiency.

**Excess (toxicity) symptoms:**
- Premature senescence of leaves may occur.

**Fertilizer sources:**
- Atmospheric deposition of S can occur downwind of large cities and industrial plants sufficient to meet S-crop requirements.
- Is becoming an increasingly occurring deficiency in many agricultural areas due to reduced atmospheric deposition and the use of low S-content fertilizers.
- Fertilizer sources are given in Table 19.1.

# 12 Micronutrients Considered Essential to Plants

The micronutrients are seven elements essential to plants and found at less than 0.01% concentration in plant dry matter. The criteria for designating an element as essential are discussed in Chapter 3.

## 12.1 TERMINOLOGY

The seven micronutrients B, Cl, Cu, Fe, Mn, Mo, and Zn have been variously identified in the past as either *trace or minor elements,* terms that are no longer used. The correct term is *micronutrient.* The micronutrients are found and required in relatively low concentrations in plants compared to the major elements (see Chapter 11, "Major Essential Plant Elements").

Micronutrient concentrations are expressed as parts per million (ppm) in this text but in SI units, the terms would be either milligrams per kilogram (mg/kg) or millimoles per kilogram (mmol/kg). Comparative values for the micronutrients in these three units are shown in this example (values selected for illustrative purposes only):

| Micronutrient | ppm | mg/kg | mmol/kg |
|---|---|---|---|
| Boron (B) | 20 | 20 | 1.85 |
| Chlorine (Cl) | 100 | 100 | 2.82 |
| Copper (Cu) | 12 | 12 | 0.19 |
| Iron (Fe) | 111 | 111 | 1.98 |
| Manganese (Mn) | 55 | 55 | 1.00 |
| Molybdenum (Mo) | 1 | 1 | 0.01 |
| Zinc (Zn) | 33 | 33 | 0.50 |

## 12.2 ESTABLISHED DATE FOR ESSENTIALITY/RESEARCHERS

Several of the micronutrients were identified as essential before the criteria for essentiality were established by Arnon and Stout (1939), criteria that have been accepted today by the research community.

For some researchers, however, these criteria may be misidentifying some elements that have been placed in an unofficial category, known as the beneficial elements. (see Chapter 13, "Elements Considered Beneficial to Plants"). The two such elements falling into this category are Ni and Si, elements that are listed at the end of this section, although they have yet to be "officially" categorized as a micronutrient.

Cobalt (Co) is an element that is required by organisms that are able to fix atmospheric nitrogen ($N_2$). Therefore, some consider Co as an essential element (would be categorized as a micronutrient based on its concentration requirement) for those legumes that are able to symbiotically fix $N_2$ that exists in the atmosphere surrounding the plant roots. Without sufficient Co, these plants would require an inorganic source of N in order to grow.

## 12.3 CONTENT AND FUNCTION

The approximate concentrations of micronutrients in mature leaf tissue generalized for various plant species from deficient to sufficient or normal and to excessive or toxic.

In general, the micronutrients Cl, Cu, Fe, and Mn are involved in various processes related to photosynthesis; therefore, their deficiency will be evident in terms of either significantly reduced plant growth and/or chlorosis symptoms. Four of the micronutrients, Cu, Fe, Mn, and Zn, are associated with various enzyme systems; Mo is specific for nitrate reductase only. Boron is the only micronutrient not specifically associated with either photosynthesis or enzyme function, but is associated with the carbohydrate chemistry and reproductive system of the plant.

## 12.4 SOIL AND PLANT SPECIES ASSOCIATIONS

The micronutrients are unique among the essential elements because their deficiency is frequently associated with a combination of crop species and soil characteristics.

A list of some of the more common crop–soil associations include the following:

| Micronutrients | Sensitive Crops | Soil Conditions for Deficiency |
|---|---|---|
| Boron (B) | Alfalfa, clover, cotton, peanut, sugar beet, cabbage and relatives, cereals, potato, tomato, celery, grapes, cucumber, sunflower, fruit trees (apple, and pear), mustard | Acid sands, soils low in organic matter, overlimed soils, organic soils |
| Copper (Cu) | Corn, onions, small grains, watermelon, sunflower, spinach, citrus seedlings, gladiolus | Organic soils, mineral soil high in pH and organic matter |
| Iron (Fe) | Citrus, clover, pecan, sorghum, soybean, grape, several calcifuge species, rice, tobacco, clover | Leached sandy soils low in organic matter, alkaline soils, soils high in P |
| Manganese (Mn) | Alfalfa, small grains, soybean, sugar beet fruit trees (apple, cherry, citrus), legumes, potato, cabbage | Leached acid soils, neutral to alkaline soil high in organic matter |
| Molybdenum (Mo) | Alfalfa, clovers | No specific soil correlation |
| Zinc (Zn) | Corn, field beans, pecan, sorghum, legumes, grasses spinach, hops, flax, grape, citrus, soybean | Leached acid sandy soils low in organic matter, neutral to alkaline soils and/or high in P |

In addition, there are some crop plants that are uniquely sensitive to either deficiency or excess of a micronutrient:

| Micronutrient | Sensitive to Deficiency | Sensitive to Excess |
|---|---|---|
| Boron (B) | Legumes, *Brassica* (cabbage and relatives), beets, celery | |
| | Grapes, fruit trees (apple, and pears), cotton, sugar beet | Cereals, potato, tomato, cucumber, sunflower, mustard |
| Chlorine (Cl) | Cereals, celery, potato, coconut, palm, sugar beet, lettuce, carrot, cabbage | |
| | Strawberry, navy bean, fruit trees, pea, onion cereals, legumes, spinach, citrus seedlings, gladiolus | |
| Copper (Cu) | Cereals (oat), sunflower, spinach, alfalfa, onion, watermelon | |
| Iron (Fe) | Fruit trees (citrus), grape, several calcifuge species, pecan, sorghum, soybean, sugar beet | Rice, tobacco |
| Manganese (Mn) | Cereals (oat), legumes, fruit trees (apple, cherries, citrus), soybean, sugar beet | Cereals, legumes, potato, cabbage |
| Molybdenum (Mo) | *Brassica* (cabbage and relatives), legumes | Cereals, pea, green bean |
| Zinc (Zn) | Cereals (corn), legumes, grasses, flax, grape, fruit trees (citrus), soybean, fieldbean, pecan | Cereals, spinach |

## 12.5  MICRONUTRIENT CHARACTERISTICS

The characteristics of the micronutrients, their approximate concentrations in plants and the environment, and plant function are given in Table 12.1.

## 12.6  MICRONUTRIENT PROPERTIES

### 12.6.1  BORON (B)

**Established date for essentiality/researchers:**

- 1928 / Sommer and Lipman

**Functions in plants:**

- Believed to be important in the synthesis of one of the bases for RNA (uracil) formation.
- In cellular activities (e.g., division, differentiation, maturation, respiration, growth, etc.).
- Has long been associated with pollen germination and growth, improving the stability of pollen tubes.
- Relatively immobile in plants.
- Transported primarily in the xylem.

**TABLE 12.1**

**Micronutrients Essential for Growth and Their Average Content in Material for Cultivated Higher Plants and Approximate Concentration in the Environment**

| Micronutrient | Mass Conc. (g/dry matter) | Molar Conc., mmol (kg dry matter) | No. of Atoms Relative to Mo | Conc. in Environment (mol/m³) | Examples of Functions in Cells |
|---|---|---|---|---|---|
| Chlorine (Cl) | 0.1 | 3 | $3 \times 10^3$ | 0.001 | Chloroplast photosystem II metabolism growth |
| Boron (B) | 0.02 | 2 | $2 \times 10^3$ | 0.001 | Carbohydrate chemistry, pollen germination and pollen tube development |
| Iron (Fe) | 0.01 | 2 | $2 \times 10^3$ | 0.001 | Energy transfer, proteins, co-enzyme factor prosthetic groups |
| Manganese (Mn) | 0.05 | 1 | $1 \times 10^3$ | 0.001 | Co-factor in water splitting enzyme, amino-peptidase, etc. |
| Zinc (Zn) | 0.02 | 0.3 | $3 \times 10^3$ | $7 \times 10^4$ | Enzyme co-factor, carbonic anhydrase, alkaline phosphatase, enzyme regulation |
| Copper (Cu) | 0.06 | 0.3 | $1 \times 10^3$ | $3 \times 10^4$ | Constituent of plastocyanin, ascorbic acid, oxidase, etc. |
| Molybdenum (Mo) | 0.0001 | 0.001 | 1 | $5 \times 10^4$ | Constituent of nitrate reductase |

*Source*: Porter, J.R. and D.W. Lawlor (Eds.). 1991. *Plant Growth Interaction with Nutrition and Environment*. Society for Experimental Biology. Seminar 43. Cambridge University Press, New York.

**Relative tolerance of some crops to boron:**

| High | Medium | Low |
|------|--------|-----|
| Sugar beet | Sunflower | Peanut |
| Garden beet | Cotton | Black walnut |
| Alfalfa | Radish | Navy bean |
| Gladiolus | Field peas | Pear |
| Onion | Barley | Apple |
| Turnip | Wheat | Peach |
| Cabbage | Corn | |
| Lettuce | Milo | |
| Carrot | Oats | |
| | Pumpkin | |
| | Sweet potato | |

**Content and distribution in plants:**
- B requirement is separated into three groups based on plant species (monocots, dicots, and dicots with latex systems):
  - Leaf content of monocots, 1 to 6 ppm B
  - Dicots, 20 to 70 ppm B
  - Dicots with latex system, 80 to 100 ppm B
- Tends to accumulate in leaf margins at concentrations five to ten times that found in the leaf blade; sufficiently high to result in marginal burning and death of the leaf margin.
- High Ca content in the plant creates a high B requirement.
- High K plant content accentuates the negative effect of low B tissue levels.
- Exists in the plant as the borate anion ($BO_3^{3-}$).

**Available forms for root absorption:**
- Most of the B in soil exists in organic plant and microorganism residues, released by residue decomposition, thereby becoming the major supply source for crop utilization.
- Exists in the soil solution as the borate anion ($BO_3^{3-}$).
- Recent findings suggest that B can also exist as undissociated $B(OH)_3$.
- Primary loss of B from soils is by leaching.
- Leaching is also a common technique of removing excess B from the surface soil and rooting zone.
- Total B in the soil ranges between 20 and 200 ppm.
- Amount available for plant absorption ranges from 1 to 5 ppm in the soil solution.
- Plant availability is influenced by soil water pH (see Figures 9.2 and 9.3).

**Movement in soil:**
- By mass flow and diffusion

**Causes of deficiency:**
- Low soil content

- When soil moisture levels are low for extended periods of time, or after a long period of heavy-leaching rainfall.

**Deficiency symptoms:**
- Abnormal growth of growing points (meristematic tissue) with apical growing points eventually becoming stunted, and then die.
- Auxins accumulate at growing points with leaves and stems becoming brittle.

**Excess (toxicity) symptoms:**
- Leaf tips become yellow, followed by necrosis as B accumulates in the leaf tips and margins (maybe as high at ten times than that in the leaf blade).
- Leaves eventually assume a scorched appearance and prematurely fall off.
- When the plant tissue levels exceeds 100 ppm B.

**Major antagonistic elements:**
- Ca, P, K, and N

**Major synergistic elements:**
- P and N

**Usual soil content:**
- 5 to 100 mg B/kg

**Fertilizer sources:**
- Fertilizer borate 48; $Na_2B_4O_7 \cdot 10H_2O$; 14% to 15% B
- Fertilizer borate granular; $Na_2B_4O_7 \cdot 10H_2O$; 14% B
- Foliarel; $Na_2B_8O_{13} \cdot 4H_2O$; 21% B
- Solubor; $Na_2B4O_7 \cdot 4H_2O + Na_2B_{10}O_{16} \cdot 10H_2O$; 20% B
- Borax; $Na_2B_4O_7 \cdot 10H_2O$; 11% B

**Application:**
- Either soil or foliar applied.
- Soil application ranging from 0.5 to 2.0 lbs B/A (0.56 to 2.2 kg B/ha).
- Care needed when applying because toxicity can occur from irregular soil distribution.

## 12.6.2 CHLORINE (CL)

**Established date for essentiality / researchers:**
- 1955 / Boyer, Carton, Johnson, and Stout

**Functions in plants:**
- Involved in the evolution of oxygen ($O_2$) in photosystem II in the photosynthesis process.
- Raises the cell osmotic pressure.
- Affects stomatal regulation.
- Increases the hydration of plant tissue.
- May be related to the suppression of leaf spot disease in wheat and fungus root disease in oat.

**Content and distribution in plants:**
- Leaf content of Cl ranges from low parts per million levels (20 ppm) in the dry matter to percent concentrations.
- Deficiency occurs in wheat when plant levels are less than 0.15%.

- Exists in the plant as the Cl⁻ anion.

**Available forms for root absorption:**
- Exists in the soil solution as the Cl⁻ anion.
- Moves in the soil by mass flow.
- Chloride anion (Cl⁻) competes with other anions, such as $NO_3^-$ and $SO_4^{2-}$, for uptake.

**Deficiency symptoms:**
- Chlorosis of younger leaves and wilting of the plant.
- Not common among most plants; however, for wheat and oats in some soil areas, deficiency is related to disease infestation.

**Excess (toxicity) symptoms:**
- Premature yellowing of the leaves.
- Burning of the leaf tips and margins.
- Bronzing and abscission of the leaves.
- Primarily associated with salt-affected [high-sodium chloride (NaCl) content] soils, a condition that influences the osmotic characteristics of plant roots by restricting the uptake of water and other ions.

**Fertilizer sources:**
- Cl-containing fertilizers, such as potassium chloride (KCl).
- Trace amounts of Cl found in many fertilizer materials, sufficient in most cases to satisfy the crop requirement for Cl.

## 12.6.3   COPPER (CU)

**Established date for essentiality / researchers:**
- 1925 / McHargue

**Functions in plants:**
- Constituent of the chloroplast protein plastocyanin.
- Serves as a part of the electron transport system linking photosystems I and II in the photosynthesis process.
- Participates in protein and carbohydrate metabolism and nitrogen ($N_2$) fixation.
- Is part of the enzymes that reduce both atoms of molecular oxygen ($O_2$) (cytochrome oxidase, ascorbic acid oxidase, and polyphenol oxidase).
- Is involved in the desaturation and hydroxylation of fatty acids.

**Content associated with plant conditions:**
- Requirement for most crops is quite low.
- Sufficiency range in leaves is between 3 and 7 ppm of the dry matter (see page 137).
- Toxicity range begins at 20 to 30 ppm.
- High values, 20 to 200 ppm, can be tolerated if Cu has been applied as a fungicide.

**Deficiency:**
- Not likely to occur on most mineral soils.
- Not an uncommon occurrence on both very sandy and organic soils.

**Interaction with other elements:**

- Interferes with Fe metabolism that may result in the development of an Fe deficiency.
- Interaction with Mo that may interfere with the enzymatic reduction of $NO_3$.

**Available forms for root absorption:**

- Exists in the soil primarily in complexed forms as low-molecular-weight organic compounds such as humic and fulvic acids.
- Cupric ion ($Cu^{2+}$) is present in very small quantities in the soil solution.
- Uptake rates are lower than for most other micronutrients.
- Plant availability is influenced by soil water pH (see Figures 9.2 and 9.3).

**Movement in soil and root absorption:**

- Moves in the soil solution by diffusion.
- Concentration in the soil solution is very low (<0.2 mg/kg), yet most soils have sufficient Cu to meet most crop requirements.
- Most soils are able to maintain sufficient $Cu^{2+}$ ions in soil solution even with increasing soil pH to meet crop requirements.
- With increasing organic matter content, Cu availability can be significantly reduced.

**Deficiency symptoms:**

- Reduced or stunted growth with distortion of young leaves and necrosis of the apical meristem.
- In trees, deficiency may cause white tip or bleaching of younger leaves and summer dieback.

**Excess (toxicity) symptoms:**

- Can induce Fe deficiency and leaf chlorosis.
- Root growth may be suppressed, with inhibited elongation and lateral root formation at relatively low Cu levels in the soil solution.
- Copper is about five to ten times more toxic to roots than Al and may more significantly affect root development into acid (pH <5.5) subsoils than Al.

**Major antagonistic elements:**

- Ca, P, K, and N

**Major synergistic elements:**

- Ca, P and N

**Usual soil content:**

- 5 to 100 mg Cu/kg (extreme values: 0.1 to 1,300 mg Cu/kg)

**Fertilizer sources:**

- Copper sulfate (monohydrate); $CuSO_4 \cdot H_2O$; 35% Cu
- Copper sulfate (pentahydrate); $CuSO_4 \cdot 5H_2O$; 25% Cu
- Cupric oxide; CuO; 75% Cu
- Cuprous oxide; $Cu_2O$; 89% Cu
- Cupric ammonium phosphate; $Cu(N_4)PO_4 \cdot H_2O$; 32% Cu
- Basic copper sulfates; $CuSO_4 \cdot 3Cu(OH)_2$ (general formula); 13% to 53% Cu
- Cupric chloride; $CuCl_2$; 17% Cu
- Copper chelates; $Na_2CuEDTA$, NaCuHEDTA, organically bound Cu; 5% to 7% Cu

**Application:**
- Copper can be either soil or foliar applied.

## 12.6.4 IRON (FE)

**Established date for essentiality / researchers**
- 1843 / Gris

**Functions in plant:**
- Important component in many plant enzyme systems, such as cytochrome oxidase (electron transport) and cytochrome (terminal respiration step).
- Component of protein ferredoxin and is required for $NO_3$ and $SO_4$ reduction, nitrogen $(N_2)$ assimilation, and energy (NADP) production; functions as a catalyst or part of an enzyme system associated with chlorophyll formation.
- Thought to be involved in protein synthesis and root-tip meristem growth.
- The plant and soil chemistry of Fe is highly complex, with both aspects still under intensive study.

**Content and distribution in plants:**
- Leaf content ranges from 10 to 1,000 ppm in the dry matter with sufficiency ranging from 50 to 75 ppm (see page 137).
- Total Fe may not always be correlated to sufficiency.
- 50 ppm Fe is the generally accepted critical value for most crops, with deficiency likely when total leaf Fe is less (see page 136).
- Fe extracted from the plant tissue by an organic acid (i.e., acetic acid) may be a better indicator of Fe sufficiency than total Fe in the tissue.
- Accurate determination of Fe plant content is normally not possible because extraneous sources of Fe are difficult to exclude from the determination, soil and dust contamination being the primary contributors to extraneous Fe.

**Iron function in plants:**
- Deficiency affects many crops, a common deficiency occurring on alkaline soils, and frequently referred to as *lime chlorosis.*
- The majority of plant Fe is in the ferric ion $(Fe^{3+})$ form as ferric phosphoprotein, although the ferrous ion $(Fe^{2+})$ is believed to be the metabolically active form.
- High P decreases the solubility of Fe in the plant, a P:Fe ratio of 29:1 being average for most plants.
- Potassium increases the mobility and solubility of Fe, while N accentuates Fe deficiency due to increased growth.
- The bicarbonate anion $(HCO_3^-)$ is believed to interfere with Fe uptake and translocation within the plant; high Zn can interfere with Fe metabolism, resulting in a visual symptom of Fe deficiency.

**Determining the iron status of a plant:**
- Extractable ferrous $(Fe^{2+})$ iron may be a better indicator of plant-Fe status than total Fe.
- Various extraction procedures have been proposed for diagnosing Fe deficiency, with 20 to 25 ppm being the critical extractable-Fe range.

- Most methods of determining the Fe status of the plant by means of an Fe analysis, whether total or extractable, are flawed (from the presence of soil and dust particles on tissue as well as Fe added in plant tissue processing) and pose a difficult problem in correctly assessing the Fe status of a plant.
- The indirect measurement of chlorophyll content may be the best alternative method for Fe-sufficiency diagnosis.

**Available forms for root absorption:**
- Exists in the soil as both the ferric ($Fe^{3+}$) and ferrous ($Fe^{2+}$) cations
- The $Fe^{2+}$ form, whose availability is affected by the degree of soil aeration, is thought to be the active form taken up by plants.
- Iron-sufficient plants are able to acidify the rhizosphere as well as release. Fe-complexing substances, such as siderophores, which enhance Fe availability and uptake, while Fe-inefficient plants do not exhibit similar root characteristics.
- Plant availability is influenced by soil water pH (see Figures 9.2 and 9.3).

**Iron-sensitive crops:**
- *Fruit and nut crops*: apple, apricot, avocado, banana, blueberry, brambles, cacao, cherry, coconut, coffee, grape, nuts (almond, filbert, pecan, walnut), olive, peach, pear, pineapple, plum, strawberry, tung
- *Others*: corn, rice, sorghum, soybean, sugarcane

**Movement in soil and root absorption:**
- Moves in the soil by mass flow and diffusion.
- When an Fe ion reaches the rhizosphere, it is reduced (from $Fe^{3+}$ to $Fe^{2+}$), or is released from a chelated form (although chelated Fe can also be taken into the plant as the chelate), and then absorbed.
- Cu, Mn, and Ca competitively inhibit Fe uptake, and high levels of P will also reduce Fe uptake.

**Deficiency symptoms:**
- Initially interveinal chlorosis of younger leaves.
- As the severity of the deficiency increases, chlorosis spreads to the older leaves.
- Causes of chlorosis:
  - *Weather related*: high rainfall, low temperature
  - *Soil factors*: high lime content, compaction, low soil temperature, inhibition of root growth and root activity
  - *Plant factors*: Fe-inefficient cultivars, low root:shoot ratio

**Excess (toxicity) symptoms:**
- May accumulate to several hundred ppm without toxicity symptoms.
- At a toxic level (not clearly defined), a bronzing of the leaves with tiny brown spots on the leaves will appear, a typical symptom frequently occurring with rice.

**Major antagonistic elements:**
- Ca, Mg, P, and S

**Major synergistic elements:**
- P, N, and S

**Usual soil content:**
- 0.05% to 4.0% (extreme values 10 to 80,000 mg Fe/kg)

**Fertilizer sources:**
- Ferrous ammonium phosphate; $Fe(NH_4)PO_4 \cdot H_2O$; 29% Fe
- Ferrous ammonium sulfate; $(NH_4)_2SO_4 \cdot FeSO_4 \cdot 6H_2O$; 14% Fe
- Ferrous sulfate; $FeSO_4 \cdot 7H_2O$; 19% to 21% Fe
- Ferric sulfate; $Fe(SO_4)_3 \cdot 4H_2O$; 23% Fe
- Iron chelates:
  - NaFeEDTA; 5% to 11% Fe
  - NaFeHFDTA; 5% to 9% Fe
  - NaFeEDDHA; 6% Fe
  - NaFeDTPA; 10% Fe
- Iron polyflavonoids; organically bound Fe; 9% to 10% Fe

**Application:**
- Iron sources can be either soil or foliar applied, with foliar application being the most efficient using either a solution of ferrous sulfate ($FeSO_4$) or one of the chelated (EDTA or EDDHA) forms of Fe.

## 12.6.5   MANGANESE (MN)

**Established date for essentiality/researchers:**
- 1923 / McHargue

**Functions in plants:**
- Involved in oxidation-reduction processes in the photosynthetic electron transport system.
- Essential in photosystem II for photolysis, acts as a bridge for ATP and enzyme complex phosphokinase and phosphotransferases, and activates IAA oxidases.
- Not known to interfere with the metabolism or uptake of any of the other essential elements.

**Content and distribution in plants:**
- Leaf sufficiency content of Mn ranges from 10 to 50 ppm in the dry matter in mature leaves (see page 137).
- Tissue levels will reach 200 ppm or higher (soybean, 600 ppm; cotton, 700 ppm; sweet potato, 1,380 ppm) before severe toxicity symptoms develop.
- Tends to accumulate in the leaf margins at a level five times that found in the leaf blade.

**Relative sensitivity to manganese by crop species:**
- *Low Sensitivity*: asparagus, blueberry, cotton, rye
- *Moderate Sensitivity*: alfalfa, barley, broccoli, cabbage, carrot, cauliflower, celery, clover, cucumber, corn, grass, parsnips, peppermint, sorghum, spearmint, sugar beet, tomato, turnip
- *High Sensitivity*: beans, citrus, lettuce, oats, onion, pea, peach, potato, radish, soybean, spinach, sudangrass, table beet, wheat

**Available forms for root absorption:**
- Exists in the soil solution as either $Mn^{2+}$, $Mn^{3+}$, or $Mn^{4+}$ cations and as exchangeable Mn.
- Cation $Mn^{2+}$ is the ionic form taken up by plants.
- Availability is significantly affected by soil pH, decreasing when the pH increases above 6.2 in some soils; while in other soils, the decrease may not occur until the soil water pH reaches 7.5 (see Figures 9.2 and 9.3).
- Availability can be reduced significantly by low soil temperatures.
- Soil organic matter can reduce Mn availability, decreasing its availability with an increase in organic matter content.

**Movement in soil and root absorption:**
- Primarily supplied to the plant by diffusion and root interception.
- Low soil temperature and moisture stress will reduce Mn uptake.
- Some plants may release root exudates that reduce $Mn^{4+}$ to $Mn^{2+}$, complex it, thereby increasing Mn availability to the plant.

**Deficiency symptoms:**
- For dicots, reduced or stunted growth with visual interveinal chlorosis on the younger leaves is symptomatic of Mn deficiency.
- Cereals develop gray spots on their lower leaves (*gray speck*), and legumes develop necrotic areas on their cotyledons (*marsh spot*).

**Excess (toxicity) symptoms:**
- An excess of Mn is seen on the older leaves as brown spots surrounded by a chlorotic zone or circle.
- Black specks on stone fruits, particularly for apple, and similar black specks on young bark, referred to as *measles,* is the visual evidence of high Mn content in the tissue.

**Major antagonistic elements:**
- Ca, Mg, P, and K

**Major synergistic elements:**
- Ca, Mg, P, and N

**Usual soil content:**
- 200 to 400 mg Mn/kg (extreme values: 12 to 10,000 mg Mn/kg)

**Fertilizer sources:**
- Manganese sulfate; $MnSO_4 \cdot 4H_2O$; 26% to 28% Mn
- Manganese oxide; MnO; 41% to 68% Mn
- Manganese chelate; MnEDTA; 5% to 12% Mn

**Application:**
- Best applied as a foliar spray to correct a deficiency as soil applications can be very inefficient due to soil inactivation of applied Mn.
- Row application of a P fertilizer will increase Mn availability and uptake.

### 12.6.6   MOLYBDENUM (MO)

**Established date for essentiality / researchers:**
- 1959 / Arnon and Stout

**Functions in plant:**
- Is a component of two major enzyme systems, nitrogenase and nitrate reductase, nitrogenase being involved in the conversion of nitrate ($NO_3$) to ammonium ($NH_4$).
- The requirement for Mo is reduced greatly if the primary form of nitrogen (N) available to the plant is $NH_4$.

**Content and distribution:**
- Leaf content of Mo is usually less than 1 ppm in the dry matter, due in part to the very low level of the molybdate anion ($MnO_4^{2-}$) in the soil solution.
- Normal Mo plant content ranges from 0.34 to 1.5 ppm (see page 137).
- Can be taken up in higher amounts without resulting in toxic effects to the plant.
- High Mo content (>10 ppm Mo) forage can pose a serious health hazard to cattle, particularly dairy cows, which have a sensitive Cu-to-Mo balance requirement.

**Available forms for root absorption:**
- Primary soluble soil form is the molybdate anion ($MoO_4^{2-}$), whose availability is increased tenfold for each unit increase in soil pH.
- In the soil, Mo is strongly absorbed by Fe and Al oxides, the formation of which is pH dependent.

**Movement in soil and root absorption:**
- Mass flow and diffusion equally supply Mo to roots, although mass flow supplies most of the Mo when the soil Mo level is high.
- If $NO_3$ is the primary N source, Mo uptake is higher than if $NH_4$ is equal to or greater than $NO_3$ as the source of N.
- Both P and Mg will enhance Mo uptake, while $SO_4$ will reduce the uptake of Mo.

**Deficiency symptoms:**
- Frequently resemble N deficiency symptoms: lack of dark green foliage color.
- Older and middle leaves become chlorotic first, and in some instances, leaf margins are rolled and growth and flower formation is restricted.
- Cruciferae and pulse crops have high Mo requirements (see Table 3.6). In cauliflower, the middle lamella of the cell wall is not formed completely when Mo is deficient, with only the leaf rib formed, thereby giving a *whip-tail* appearance in severe cases; *whip-tail* is a commonly used term to describe a Mo deficiency.

**Excess (toxicity) symptoms:**
- High plant Mo does not normally affect the plant, but can pose a problem for ruminant animals, particularly dairy cows, that consume plants containing 5 ppm or more Mo.

**Major antagonistic elements:**
- P, K, and S

**Major synergistic elements:**
- N and P

**Usual soil content:**
- 0.5 to 5 mg Mo/kg (extreme values: 0.1 to 80 mg Mo/kg)

**Fertilizer sources:**
- Ammonium molybdate; $(NH_4)_6Mo_7O_{24} \cdot 2H_2O$; 54% Mo
- Sodium molybdate; $Na_2MoO_4 \cdot 2H_2O$; 39% to 41% Mo
- Molybdenum trioxide; $MnO_2$; 66% Mo

**Application:**
- Best supplied by means of seed treatment

### 12.6.7 ZINC (ZN)

**Established date for essentiality / researchers:**
- 1926 / Sommer and Lipman

**Functions in plants:**
- Involved in the same enzymatic functions as Mn and Mg with only carbonic anhydrase activated by Zn.
- The relationship between P and Zn has been intensively studied as research suggests that high P can interfere with Zn metabolism as well as affect the uptake of Zn through the root.
- High Zn can induce an Fe deficiency, particularly those sensitive to Fe.

**Content and distribution:**
- In general, the leaf sufficiency range for Zn is from 15 to 50 ppm in the dry matter in mature leaves (see page 137).
- For some species, visual deficiency symptoms will not appear until the Zn content is as low as 12 ppm.
- For the majority of crops, 15 ppm Zn in the leaves is considered the *critical value* (see page 136).
- A small variation in Zn content, as little as 1 to 2 ppm at the critical level, may be sufficient to establish either deficiency or sufficiency.
- Some plants can accumulate considerable quantities of Zn (several 100 ppm) without harm to the plant.

**Available forms for root absorption:**
- Exists in the soil solution as the $Zn^{2+}$ cation, as exchangeable Zn, and as organically complexed Zn.
- Availability is affected by soil pH, decreasing with increasing pH.
- Availability can also be reduced when the available soil P level is very high.
- Plant availability is influenced by soil water pH (see Figures 9.2 and 9.3).

**Movement in soil and root absorption:**
- Brought into contact with plant roots by mass flow and diffusion, with diffusion being the primary delivery mechanism.
- $Cu^{2+}$ and other cations, such as $NH_4^+$, will inhibit root Zn uptake.
- P appears to inhibit translocation rather than directly inhibiting uptake.
- Efficiency of Zn uptake seems to be enhanced by a reduction in pH of the rhizosphere, with those plant species that can readily reduce the pH being less affected by low soil Zn than those that cannot.

**Deficiency symptoms:**
- Appears as a chlorosis in the interveinal areas of new leaves, producing a banding appearance.
- With increasing severity of the deficiency, leaf and plant growth become stunted (rosette), and leaves die and fall off the plant.
- At branch terminals of fruit and nut trees, rosetting occurs with considerable die-back of the branches.

**Excess (toxicity) symptoms:**
- Plants particularly sensitive to Fe (see page 85) will become chlorotic when Zn levels are abnormally high (>100 ppm).
- There are some plant species that can tolerate relatively high Zn contents (100 to 250 ppm) without any significant effect on plant growth and yield.

**Major antagonistic elements:**
- Ca, Mg, and P

**Major synergistic elements:**
- Ca, Mg, and P

**Usual soil content:**
- 10 to 500 mg Zn/kg (extreme values: 4 to 10,000 mg Zn/kg)

**Fertilizer sources:**
- Zinc sulfate; $ZnSO_4 \cdot 7H_2O$; 35% Zn
- Zinc oxide; ZnO; 78% to 80% Zn
- Zinc chelates:
  - $Na_2ZnEDTA$; 14% Zn
  - NaZnTA; 13% Zn
  - NaZnHEDTA; 9% Zn
- Zinc polyflavonoids; organically bound Zn; 10% Zn

**Application:**
- Can be applied to the plant by means of both soil and foliar applications.

## 12.7  POSSIBLE ADDITIONAL ESSENTIAL MICRONUTRIENTS

As discussed at the beginning of this section, the author is adding two elements, not yet identified as micronutrients to this discussion

### 12.7.1  NICKEL (NI)

**Established date for essentiality / researchers:**
- 1987 / Brown, Welsh and Cary

**Functions in plants:**
- Component of plant urease.
- Benefits growth of N-fixing plant species.
- Barley seeds will not germinate when deficient in Ni.

**Content and distribution:**
- Adequate range between 0.084 and 0.22 ppm in leaf dry matter, <0.1 ppm for grain crops, and 0.2 ppm for legumes.

- Toxicity occurs when the Ni content in plant leaf dry matter exceeds 10 ppm for Ni-sensitive plants and up to 50 ppm for Ni-tolerant plants.

**Available forms for root absorption:**

- Is available and absorbed as the $Ni^{2+}$ cation.

**Movement in soil and root absorption:**

- Brought into contact with plant roots by mass flow and diffusion, with diffusion being the primary delivery mechanism.
- Uptake can be partially inhibited by high levels of $Cu^{2+}$ and $Zn^{2+}$ cations in the soil solution.
- Once absorbed, Ni is readily redistributed in the plant.

**Deficiency symptoms:**

- Significant reduction in shoot growth for barley, oat, and wheat plants.
- Reduced barley grain germination.
- Leaf tip burn due to the accumulation of toxic levels of urea.
- Reduced and delayed leaf expansion.

**Excess (toxicity) symptoms:**

- Leaf symptoms may be similar to Fe-deficiency chlorosis.
- Sewage sludge applied to a cropland soil may result in Ni toxicity.

**Fertilizer sources:**

- Exists in most soils at sufficient levels to meet most crop requirements; therefore Ni is not generally applied as or with fertilizer.
- Availability is significantly reduced on high pH and lime content soils where soil fixation is likely to occur.
- Major source of Ni is sewage sludge.

## 12.7.2   SILICON (SI)

**Established date for essentiality / researchers:**

- 1987 / Brown, Welsh, and Cary

**Functions in plants:**

- Maintains stalk strength in small grains, particularly rice by preventing stalk lodging.
- Provides cellular protection, limiting the invasion of fungi into plant cells.

**Content and distribution:**

- Deficiency concentration not established; concentration under natural conditions may be at percentage levels in plant dry matter.

**Available forms for root absorption:**

- Probably brought into contact with plant roots by mass flow and diffusion, with diffusion being the primary delivery mechanism; exact mechanisms for movement within the soil not well known.

**Deficiency symptoms:**

- Plants easily lodge, lack stalk strength.

**Excess (toxicity) symptoms:**

- Not known

# 13 Elements Considered Beneficial to Plants

Although no such category has been officially established, many believe that more than the sixteen currently categorized "essential elements" (see Chapter 11, "Major Essential Plant Elements," and Chapter 12, "Micronutrients Essential to Plants") must be present in the plant to ensure maximum growth.

## 13.1 THE A TO Z NUTRIENT SOLUTION

In earlier times, researchers growing plants hydroponically formulated an A to Z solution that was added to the formulated nutrient solution in order to ensure that those elements commonly found in the environment (soil, plants, and water) were also present in the nutrient solution. The A to Z solution included the following twenty elements without any concentration levels given:

| | | |
|---|---|---|
| Aluminum (Al) | Arsenic (As) | |
| Barium (Ba) | Bismuth (Bi) | Bromine (Br) |
| Cadmium (Cd) | Chromium (Cr) | Cobalt (Co) |
| Fluorine (F) | Iodine (I) | Lithium (Li) |
| Lead (Pb) | Mercury (Hg) | Nickel (Ni) |
| Rubidium (Rb) | Selenium (Se) | Strontium (Sr |
| Tin (Sn) | Titanium (Ti) | Vanadium (V) |

Today, solutions such as the A to Z nutrient solutions are not recommended for use.

## 13.2 ELEMENTS ESSENTIAL FOR ANIMALS

Eight of the twenty elements included in the A to Z micronutrient solution are considered essential and/or beneficial for animals, those elements being

| | |
|---|---|
| Arsenic (As) | Iodine (I) |
| Cobalt (Co) | Nickel (Ni) |
| Chromium (Cr) | Selenium (Se) |
| Fluorine (F) | Vanadium (V) |

Many feel that these eight elements recognized as essential or beneficial for animals but not currently for plants (see Table 13.1) are good candidates for investigating their possible essentiality. Those who wish to explore the potential for discovery

**TABLE 13.1**
**Essentiality of Micronutrients for Plants and Animals**

| Element | Plants | Animals |
|---|---|---|
| Arsenic (As) | No | Yes |
| Boron (B) | Yes | No |
| Chromium (Cr) | No | Yes |
| Cobalt (Co) | No | Yes |
| Copper (Cu) | Yes | Yes |
| Fluorine (F) | No | Yes |
| Iodine (I) | No | Yes |
| Iron (Fe) | Yes | Yes |
| Lithium (Li) | No | Yes? |
| Manganese (Mn) | Yes | Yes |
| Molybdenum (Mo) | Yes | Yes |
| Nickel (Ni) | Yes? | Yes |
| Selenium (Se) | No | Yes |
| Silicon (Si) | Yes? | No |
| Zinc (Zn) | Yes | Yes |

of additional elements that may prove essential for both animals and plants will find the articles by Mertz (1981), Asher (1991), and Pais (1992) interesting reading.

## 13.3   BASIS FOR ESSENTIALITY FOR BENEFICIAL ELEMENTS

Some plant physiologists feel that the criteria for essentiality established by Arnon and Stout (1939, see page 18) could preclude the addition of other elements, as the present sixteen identified as essential include most of the elements found in substantial quantities in plant tissues. However, there may be other elements that have yet to be proved essential, as their requirement is at such low concentrations in plant tissues that it will take considerable, sophisticated analytical skills to uncover them, or their presence in the environment will require special skills to remove them from the rooting medium in order to create a deficiency. This was the case for Cl, the last of the essential elements to be so defined (see page 88). The question is: What are the elements likely to be added to the list as being essential, and where is the best place to begin? To complicate matters, plant response to some elements is species related as not all plants respond equally to a particular element (Pallas and Jones, 1978).

It should be remembered that since the beginning of time, plants have been growing in soils that contain all currently known elements. Those elements, when existing in the soil solution as ions, can be taken into the plant by root absorption. This means that plants will contain most of, if not all those elements found in the soil solution (element concentration in the solid phase does not equate to that found in the soil solution as the physiochemical properties determine that concentration in the soil

**TABLE 13.2**
**Trace Element Content of Markert's Reference Plant**

| Trace Element | mg/kg | Trace Element | mg/kg |
|---|---|---|---|
| Antimony (Sb) | 0.1 | Iodine (I) | 3.0 |
| Arsenic (As) | 0.1 | Lead (Pb) | 1.0 |
| Barium (Ba) | 40 | Mercury (Hg) | 0.1 |
| Beryllium (Be) | 0.001 | Nickel (Hi) | 1.5 |
| Bismuth (Bi) | 0.01 | Selenium (Se) | 0.02 |
| Bromine (Br) | 4.0 | Silver (Ag) | 0.2 |
| Cadmium (Cd) | 0.05 | Strontium (Sr) | 50 |
| Cerium (Ce) | 0.5 | Thallium (Ti) | 0.05 |
| Cesium (Cs) | 0.2 | Tin (Sn) | 0.2 |
| Chromium (Cr) | 1.5 | Titanium (Ti) | 5.0 |
| Fluorine (F) | 2.0 | Tungsten (W) | 0.2 |
| Gallium (Ga) | 0.1 | Uranium (U) | 0.01 |
| Gold (Au) | 0.001 | Vanadium (V) | 0.4 |

*Note:* No data from typical accumulator and/or rejecter plants.

*Source:* Markert, B. 1994. Trace element content of "Reference Plant." In D.C. Adriano, Z.S. Chen, and S.S. Yang (Eds.), *Biochemistry of Trace Elements*. Science and Technology Letters, Northwood, NY.

solution). Markert (1994) defined what he called the "Reference Plant Composition" that included twenty-six elements that are not essential but are found in plants at easily detectable concentrations (Table 13.2). Some of these elements would be classified as trace elements (see Chapter 16, "Soil Testing") because those found in the plant's dry matter exist at low concentrations. This designation, however, can lead to some confusion because the term trace elements was once used to identify what is defined today as micronutrients (see Chapter 12, "Micronutrients Essential to Plants"). Some of these elements exist at fairly high concentrations in the plant, depending on the level of their availability in the soil solution. Kabata-Pendias and Pendias (2000) have given the approximate concentration of eighteen nonessential elements found in plant leaf tissue, giving their range of sufficiency to toxicity to excess (Table 13.3). The question is: Which of the elements listed in Tables 13.2 and 13.3, irrespective of their observed concentration in plants, would contribute positively or negatively to plant growth? A colleague and this author (Pallas and Jones, 1978) found that platinum (Pt) at very low level [0.057 mg/L (ppm)] in a hydroponic nutrient solution stimulated plant growth for some plant species, but then at a higher level [i.e., 0.57 mg/L (ppm)] reduced growth for all species. The growth effects at the low level of Pt in solution varied considerably among the nine plant species included in the study (no response—radish and turnip; positive response—snapbean, cauliflower, corn, peas, and tomato; negative response—broccoli and pepper). It is the stimulatory effect of an element that must be investigated for those elements found, either in soils or soilless media, and/or when added to a hydroponic nutrient solution, would benefit plant growth.

**TABLE 13.3**

**Approximate Concentrations of Micronutrients and Trace Elements in Mature Leaf Tissue**

| Trace Element | Normal | Toxic | Excessive |
|---|---|---|---|
| | | (mg/kg dry wt) | |
| Antimony (Sb) | — | 7–50 | 150 |
| Arsenic (As) | — | 1–1.7 | 5–20 |
| Barium (Ba) | — | — | 500 |
| Beryllium (Be) | — | <1–7 | 10–50 |
| Boron (B) | 5–30 | 10–200 | 5–200 |
| Cadmium (Cd) | — | 0.05–0.2 | 5–30 |
| Chromium (Cr) | — | 0.1–0.5 | 5–30 |
| Cobalt (Co) | — | 0.02–1 | 15–50 |
| Copper (Cu) | 2–5 | 5–30 | 2–100 |
| Fluorine F) | — | 5–30 | 50–500 |
| Lead (Pb) | — | 5–10 | 30–300 |
| Lithium (Li) | — | 3 | 5–50 |
| Manganese (Mn) | 15–25 | 20–300 | 300–500 |
| Molybdenum (Mo) | 0.1–0.3 | 0.211 | 0–50 |
| Nickel (Ni) | — | 0.1–5 | 10–100 |
| Selenium (Se) | — | 0.001–2 | 5–30 |
| Silver (Ag) | — | 0.5 | 5–10 |
| Thallium (Tl) | — | — | 20 |
| Tin (Sn) | — | — | 60 |
| Titanium (Ti) | 0.2–0.5 | 0.5–2.0 | 50–200 |
| Vanadium (V) | — | 0.2–1.5 | 5–10 |
| Zinc (Zn) | 10–20 | 27–150 | 100–400 |
| Zirconium (Zr) | 0.2–0.5 | 0.5–2.0 | 15 |

*Source:* Kabata-Pendias, A. and H. Pendias. 2000. *Trace Elements in Soils and Plants, 3rd ed.* CRC Press, Boca Raton, FL.

## 13.4  POTENTIAL ESSENTIAL ELEMENTS

Four elements—Co, Ni, Si, and V—have been identified as to their potential essentiality for plants, as considerable research has been devoted to each of these elements, and some investigators feel that they are important elements for sustaining vigorous plant growth.

### 13.4.1  COBALT (CO)

Cobalt is required indirectly by leguminous plants because this element is essential for the *Rhizobium* bacteria, which live symbiotically in the roots, fixing atmospheric nitrogen ($N_2$) and providing the host plant most, if not all, of its needed N. Without Co, the *Rhizobium* bacteria are inactive and the legume plant would then require an inorganic source of N existent as ions (either $NO_3^-$ and/or $NH_4^+$) in the soil solution. It is not clear whether the plant itself also requires Co to carry out this specific biochemical process. The irony of this relationship between *Rhizobium* bacteria and leguminous plants is that in the absence of sufficient inorganic N in the soil, and therefore wholly dependent on $N_2$ fixed by the *Rhizobium* bacteria, the plant will be deficient in N, cease to grow, and eventually die if Co is not present.

### 13.4.2   SILICON (SI)

Plants that are soil-grown can contain substantial quantities of Si, equal in concentration (% levels in the dry matter) to that of some of the major essential elements (see Chapter 11, "Major Essential Plant Elements"). Most of the Si absorbed [plants can readily absorb silicic acid $(H_4SiO_4)$] is deposited in the plants as amorphous silica, $SiO_2 \cdot nH_2O$, or as opals. Epstein (1994) has identified six roles of Si in plants, both physiological and morphological. Reviewing 151 past nutrient solution formulations, Hewitt (1966) found that only a few included the element Si. Epstein (1994) recommends that Si as sodium silicate $(Na_2SiO_3)$ be included in a Hoagland nutrient solution formulation at 0.25 mM (see page 199). Morgan (2000) reported that in hydroponic trials, yield improvements for lettuce and bean crops occurred when the Si content in the nutrient solution was at 140 ppm. Recent studies with greenhouse-grown tomato and cucumber have shown that without adequate Si, plants are less vigorous and unusually susceptible to fungus disease attack (Belanger et al., 1995). Best growth was obtained when the nutrient solution contained 100 mg/L (ppm) silicic acid $(H_4SiO_4)$. Other common reagent forms of Si added to a nutrient solution are either Na or K silicates, as both are water-soluble compounds, while silicic acid $(H_4SiO_4)$ is only partially water soluble.

Silicon has been found to be required to maintain stalk strength in rice (Takahashi and Miyake, 1977), and other small grains (Takahashi et al., 1990). In the absence of adequate Si, these grain plants will not grow upright, tending to lodge, and resulting in significant grain loss. The problem of lodging has been observed primarily in paddy rice, where soil conditions may affect Si availability and uptake.

There may be confusion about this element as frequently the element silicon (Si) is improperly referred to as silica, $SiO_2$, which is an insoluble compound.

### 13.4.3   NICKEL (NI)

Nickel is considered an essential element for both legumes and small grains (e.g., barley) as Brown, Welsh, and Cary (1987) have shown that its deficiency meets one of the requirements for essentiality established by Arnon and Stout (1939) as seed produced on a Ni-deficient plant will not germinate. Nickel is a component of the enzyme urease, and plants deficient in Ni have high levels of urea in their leaves. Nickel-deficient plants are slow growing, and for barley, viable grain is not produced. It is recommended that a nutrient solution contain a Ni concentration of at least 0.057 mg/L (ppm) in order to satisfy the plant requirement for this element, although its requirement for other than grain crops has not been established. Seed viability associated with Ni efficiency for plants species other than barley has yet to be demonstrated.

## 13.5 NEW BENEFICIAL ELEMENTS

Morgan (2000) has identified what she calls "new" beneficial elements, those being other than Co, Si, Ni, and V, as:

| | |
|---|---|
| Sodium (Na) | Rubidium (Rb) |
| Strontium (Sr) | Lithium (Li) |
| Aluminum (Al) | Selenium (Se) |
| Iodine (I) | Titanium (Ti) |
| Silver (Ag) | |

All of these elements, other than Na and Ag, were not included in the A–Z Micronutrient Solution described above. Morgan (2000) has described the roles of these twelve elements in plants (Table 13.4).

## 13.6 ELEMENT SUBSTITUTION

There is considerable evidence that some nonessential elements can partially substitute for an essential element, such as Na for K, Rb for K, Sr for Ca, and V for Mo. These partial substitutions may be beneficial to plants in situations where an essential element is at a marginally sufficient concentration. For some plant species, this partial substitution may be highly beneficial to the plant, for example the substitution of Na for K in sugar beets. Vanadium seems capable of substituting for Mo in the N metabolism of plants, with no independent role clearly established for V. If Mo is at its sufficiency level [its requirement is extremely low (see Tables 13.1 and 13.2)] in the plant, V presence and availability are of no consequence.

Despite considerable speculation, it is not known exactly how or why such substitutions take place, although similarity in elemental characteristics (atomic size and valance) may be the primary factors.

## 13.7 FORM OF RESPONSE

There are two kinds of response due to the presence of those elements that have a beneficial effect:

1. Direct effect that relates specifically to that element
2. Enhancement of growth by means of substitution for an essential element

An element that fits the first effect is Si, which enhances the growth and appearance of rice. Without Si rice plants lodge readily and lack stem stiffness. Therefore, for the successful culture of rice, Si must be present in sufficient concentrations to

## TABLE 13.4
## The "New" Beneficial Elements and Their Roles

| Element | Role |
|---------|------|
| Silicon (Si) | Available as silicic acid ($H_4SiO_4$), which is slightly soluble; moves in the plant in the transpiration stream in the xylem; important role in growth, mineral nutrition, mechanical support; resistance to fungal diseases |
| Sodium (Na) | Can be a replacement for K in some plants, such as spinach and sugar beet, while small quantities have increased tomato yields; an element that can be beneficial at low concentrations and detrimental at high concentrations |
| Cobalt (Co) | Accelerates pollen germination; elevates the protein content of legumes; contributes to the maximum occupation of the leaf surface by chloroplasts and pigments; symbiotic $N_2$-fixation by legumes |
| Vanadium (V) | Complements and enhances the functioning of Mo; evident V and Mo in the $N_2$-fixation process; contributes to the initial stages of seed germination |
| Lithium (Li) | Some plants can accumulate Li to high concentrations; may affect the transport of sugars from leaves to roots in sugar beets; increases chlorophyll content of potato and pepper plants |
| Rubidium (Rb) | May partially substitute for K, when P and $NH_4$-N are high in the plant; enhances yield of fertile soil; may play a role in sugar beet plant by enhancing yield and sugar content |
| Strontium (Sr) | May partially replace Ca when Ca requirement is high |
| Aluminum (Al) | May be beneficial to plants that accumulate Al; traces found in DNA and RNA |
| Selenium (Se) | Stimulates growth for high Se accumulator plants; can replace S in S-amino acids in wheat |
| Iodine (I) | Stimulates growth of plants in I-deficient soils; stimulates the synthesis of cellulose and lignification of stem tissue; increases concentration of ascorbic acid; seems to increase the salt tolerance of plants by lowering Cl uptake |
| Titanium (Ti) | May play a role in photosynthesis and $N_2$-fixation; increases chlorophyll content of tomato leaves; increases yield, fruit ripening, and sugar content of fruit; may be essential for plants but not found so because Ti is almost impossible to remove from the environment |
| Silver (Ag) | Induces production of male flowers on female plants; blocks the production of ethylene; cut flower life can be enhanced by pretreatment with Ag compounds |

*Source:* Morgan, L. 2000. Beneficial elements for hydroponics: A new look at plant nutrition. *The Growing Edge* 11(3):40–51.

ensure high plant performance. Two other effects of Si are disease resistance—as the presence of Si in the plant is frequently associated with fungus resistance—and enhancement of Al and Fe tolerance by plants.

There have been experiments conducted that suggest that Ti can enhance plant growth when supplied in a chelated form, which allows Ti to be readily absorbed and utilized by the plant.

An example of the second effect for a beneficial element is Na that has been found to enhance plant growth and performance for some crops because Na acts as a partial substitute for K. This enhancement/substitution by Na may be species-related, and therefore not applicable to all types of plants. A similar substitution has been suggested for Mo by V.

## 13.8 SUMMARY

Evidence exists that there are beneficial effects for certain elements that are not currently recognized as essential for plants, and this may be sufficient to alert growers that the use of, for example, pure reagents and purified water for making a nutrient solution may not be the best practice. The presence of small quantities of elemental impurities may be desirable. What consideration should be given to specifically include those elements considered beneficial in the nutrient solution formula, such as including the use of the A-Z micronutrient solution (see Chapter 25, "Organic Farming/Gardening"), is questionable because the requirement to include the right amount as well as the added cost for the reagents may far exceed any derived benefit. There is also the danger that a particular element, if in excess, can adversely affect plant growth. These situations present real problems for plant physiologists as well as growers when using a nutrient solution as the only source of supply for the elements, whether essential or only considered beneficial. Hewitt (1966) did an extensive study on what elements and their concentrations exist in sand, other growing media, water from various sources, and reagents used to make a nutrient solution. Although some of this data might not apply today, particularly for contaminant elements found in currently used reagents, Hewitt's findings point out that elements do exist in most of the items used in hydroponic equipment, water, reagents, etc.

In a soil or soilless growing medium, whether inorganic or organic, many of these elements exist naturally, thereby precluding the necessity for adding them to the rooting medium as an individual element or in a fertilizer formulation. In addition, many of these same elements exist in naturally occurring materials used as liming materials (see Chapter 19, "Inorganic Fertilizers and their Properties") and as fertilizer sources (see Chapter 20, Organic Fertilizers and Their Properties") for the major elements.

Morgan (2000) states that "we have a lot to discover about the role of many elements present in plants. Most trace elements may never be considered 'essential' but they may prove, or have been proven, to be highly beneficial certainly makes them worth consideration."

Much of the interest in the so-called beneficial elements may be academic as their presence in plants could be only consequential. It also may be that their presence in plants may have no observed effect on plant growth unless conditions exist that would make that element a limiting factor. There is also the danger that their presence in plants may be detrimental, and therefore should be excluded from the rooting media and nutrient solution rather than included. Hayden (2003) warns of the use of some organic materials, such as sawdust, rice straw, and composted garbage that can contain substances harmful to plants, both organic substances, such as pesticides and herbicides, and heavy metals. Heavy metals (As, Cd, Cr, Pb, Hg) may also be present in inorganic fertilizers at levels that can harm plants. The Association of American Plant Food Control Officials (AAP-FCO) has developed and published recommended standards for "risk-based acceptable concentrations" (RBC's) on its website (http://www.aapfco.org).

# 14 Elements Considered Toxic to Plants

An official category, "toxic elements," does not exist with some elements depending on their concentration in the rooting medium or the plant itself; these toxic elements can adversely affect plants, and are therefore designated as "toxic."

## 14.1 INTRODUCTION

It is known that some elements commonly found in soils, even those that are "essential," can adversely affect plants at high levels of availability in the rooting medium as well as their presence in plant tissues. For example, the micronutrients B, Cu, Mn, and Zn can be "toxic" to plants when present in high concentration in the soil solution and hydroponic nutrient solutions (see Chapter 12, "Micronutrients Considered Essential to Plants," and 23, "Hydroponics"). High soil content does not always equate to high availability as elemental root adsorption occurs from the soil solution (see Chapter 4). Therefore, soil factors, such as texture, cation exchange capacity, organic matter and P content, and pH, will determine what portion of an element in the soil will exist in the soil solution at any particular time and at what concentration. For many of these elements, there exists an equilibrium between that in the solid phase (both organic and inorganic) of the soil and that found in solution; therefore, those factors affecting this equilibrium will also determine availability, and therefore the possibility of creating plant "toxicity" when absorbed (see Chapter 2). Plants themselves have the ability to adjust to metal-element excesses. Some of the possible mechanisms involved in tolerance include

- Selective uptake of ions
- Decreased permeability of membranes or other differences in the structure and function of membranes
- Immobilization of ions in roots, foliage, and seeds
- Removal of ions from metabolism by deposition (storage) in fixed and/or insoluble forms in various organs and organelles
- Alterations in metabolic patterns—increased enzyme system that is inhibited, or increased antagonistic metabolite, or reduced metabolic pathway by passing an inhibited site
- Adaptation to toxic metal replacement of a physiological metal in an enzyme
- Release of ions from plants by leaching from foliage, guttation, leaf shedding, and excretion from roots

## 14.2 THE NATURE OF ELEMENTAL TOXICITY

The toxicity effect may be direct, that is, the element itself directly impacts the plant, or the effect may be indirect by reducing the availability of another element or by interfering with a normal physiological process within the plant. An example of a direct toxic-effect element is B, which can significantly reduce the growth of a crop, such as corn, when applied at rates that would be required for a high B requirement crop, such as peanut or cotton. The carryover of Cu from the long-term use of Cu-based fungicides in orchards and vineyards poses significant problems for subsequent crops. Another direct toxic-effect element is the micronutrient Mn, which can be elevated to toxic concentrations in the soil solution when the pH of a mineral soil is very acidic (i.e., pH < 5.5).

An example of an indirect effect element is the micronutrient Zn. If present at high concentrations in the soil solution and/or plant, Zn will interfere with normal Fe metabolism in the plant, resulting in the development of typical Fe deficiency symptoms.

Another essential micronutrient that can indirectly affect plant growth is Cl, which normally exists in the soil solution as the chloride ion ($Cl^-$). The chloride ion effect is usually due to the presence of salt (NaCl) at concentrations that restrict water and nutrient uptake by plant roots. *Salinity* is a major problem in many areas of the world, the result of the continued use of high salt-containing irrigation water.

The most common nonessential element that is toxic to plants is Al, which can reach toxic levels in the soil solution due to soil acidity. The toxic plant symptoms can take various forms. Frequently, plants are stunted, root development is impaired, and plants may appear as P-deficient because Al interferes with the uptake of P by the roots.

It should be remembered that many elemental toxicities on cropland soils are frequently man-made problems associated with the past and present, including both antagonistic and synergistic effects. It has been suggested by some that one of the major roles of Ca in the plant is to counter the toxicity effect of heavy metals.

## 14.3 ALUMINUM AND COPPER TOXICITY

The elements Al and Cu, when present in the soil solution in relatively high concentrations, can affect plant growth and development due primarily to their toxic effect on root development and function without translocation to the upper portions of the plant.

Aluminum, constituting 50% of the elemental content of mineral soils, when brought into solution can be toxic to plants. Aluminum availability is defined by its concentration in the soil solution, increasing with decreasing soil pH. The ability of Al to be adsorbed by plant roots increases under anaerobic soil conditions and as a result of physical injury to the root. High Al concentrations found in leaf tissue, that not due to soil contamination (see page 132), can be a good indicator of the aeration status in the soil, and/or whether plant roots have been injured mechanically or by the activity of soil microorganisms. Aluminum is easily root absorbed and translocated within the plant at the initial stages of plant growth, but then sharply declines with advancing maturity. The absorption of Al by the plant root is also affected by the concentration of other elements in the rhizosphere (see page 35), such as P and Ca, their presence reducing Al root absorption. Aluminum toxicity is easily controlled by maintaining the soil pH within the optimum range (see page 57), thereby minimizing the occurrence of anaerobic soil conditions and thus preventing root injury.

Copper is an element that is toxic to plant roots at elevated concentrations in the soil solution, occurring primarily in the subsoil of residual soils when the subsoil water pH is less than 5.4 (see Figure 9.2). At comparable concentrations, Cu is more toxic to plant roots than Al, and for some plant species that are particularly sensitive to Cu. This element was a constituent of fungicides previously used for leaf disease control in orchard crops. When orchard soils were converted over to row crop production, depending on the crop and soil fertility status, the residue effect from past Cu-based fungicide use may be evident (poor plant growth) for some years. The availability of Cu in the soil can be significantly reduced by maintaining the water soil pH above 6.0.

## 14.4   OTHER ELEMENTS

Land disposal of waste products carries with it the possibility of elemental toxicity, particularly for those elements that are identified as heavy metals (see page 112). Animal manures applied to soils as a source for fertilizer elements and/or a means of disposal may include elements that can become toxic to plants due to their accumulation with repeated use. Zinc toxicity can occur with the use of some sources of sewage sludge. In addition, Cd, Cr, Hg, Cu, Pb, and Ni are elements commonly found in sewage sludge, their presence and concentration depending on their source and sewage treatment procedures. Having these substances analyzed for their elemental content is important in order to determine what elements and at what concentrations they are being applied to the soil. There are federal and some state regulations that define the load limits that apply to the use of these substances based on their heavy metal content [maximum acceptable concentration (MAC)], specifying rates [maximum application limit (MAL)], and frequency of application required to prevent plant effects, as well as avoiding the movement of these elements into the food chain. Some of the load limits are specific for particular soils based on their physiochemical properties, such as organic matter content, cation exchange capacity, texture, and pH. Therefore, the requirements that specify their use can be correlated with these particular soil properties.

## 14.5   PLANT SPECIES FACTOR

Root absorption and toxicity effects are related to plant species, as some plant species are known as accumulator plants (see page 116), and for those plant species that have unique root characteristics that affect their ability to "forage" the soil and/or penetrate deep into the soil profile. Those plant species sensitive to a deficiency of an essential element will be equally sensitive to its excess. An excellent example is soybean, which has a fairly narrow range between deficiency and toxicity for the element Mn. Also, B may fall into this same category for those species that have a fairly low "sufficiency value" for this micronutrient. For some plant species, the ease of movement within the plant for some elements results in their accumulation in a particular part of the plant, such as the seed. An example of such a situation is the accumulation of Cd in sunflower seed.

## 14.6   THE HEAVY METALS

There is a class of elements defined as heavy metals, elements whose atomic weights are greater than 50. The micronutrients Cu, Fe, Mn, and Zn are heavy metals. From an environmental aspect, the elements Cd, Cr, Hg, and Pb are heavy metals that are beginning to be found in ever-increasing concentrations in soils as a result of cropland disposal of industrial waste products, sewage sludge, and some forms of animal manures where elements such as Cu are used to control disease. Plant toxicity is usually not of primary concern unless the soil levels become excessive, but these heavy metals can be absorbed by plants and then introduced into the food chain by consumed plant products. Therefore by this introduction, concerns regarding human and animal health are raised. As stated previously, cropland application of waste products is being controlled by both federal and state regulations with changing parameters depending on the character of the waste product, cropland disposal procedures, and soil physiochemical properties (see Chapter 8, "Physiochemical Properties of Soil").

There are so-called "accumulator" plants (see page 116) that can concentrate an element in a portion of the plant, a portion that may be consumed by animals and humans. Knowing the plant species and the heavy metal element, avoidance procedures can be used to ensure that such an introductions does not occur.

# 15 Trace Elements Found in Plants

The word combination "trace element" was once used to identify those elements that today are called "micronutrients"—B, Cl, Cu, Fe, Mn, Mo, and Zn—elements essential for plants in order to complete their life cycle (see Chapter 12, "Micronutrients Considered Essential to Plants"). Because these elements are found in relatively low concentrations (i.e., less than 0.01% of dry weight in plant tissue), the designation "trace" was considered appropriate at the time of their discovery.

## 15.1 DEFINITION

Today, the use of the words "trace elements" signifies those elements found in plant tissues at relatively low concentrations, which can range from 0.01 to 0.001% or less of the dry weight, but are not considered, or as yet to be found essential because they do not meet the criteria for essentiality as set forth by Arnon and Stout (1939). Some of these elements are of particular interest due to their effect on plant growth if found in relatively high concentrations, and particularly if present in the harvested product that may result in adverse effects when consumed by animals or humans. A number of these elements are also considered "toxic," a topic that is discussed in Chapter 14, "Elements Considered Toxic to Plants."

## 15.2 ELEMENTS CATEGORIZED AS TRACE ELEMENTS

These elements are not specific, in that the assigned term "trace element" would not immediately identify the element. Elements that fall into this category include:

- Aluminum (Al): a major soil constituent
- Arsenic (As): essential for animals
- Cadmium (Cd): high soil content is usually due to human activity
- Chromium (Cr): essential for animals
- Cobalt (Co): required element for symbiotic $N_2$-fixation bacteria associated with leguminous plants, essential for animals
- Fluorine (F): essential for animals
- Lead (Pb): widely present from the past use of leaded gasoline
- Mercury (Hg): high soil levels are usually due to human activity
- Nickel (Ni): considered by some as an essential micronutrient
- Selenium (Se): essential for animals
- Silicon (Si): considered essential for grain crops, particularly rice
- Strontium (Sr): some suggest that this element will partially substitute for Ca

- Titanium (Ti): a major constituent in soils
- Vanadium (V): will substitute for molybdenum (Mo), is essential for animals

Some of these trace elements occur naturally in soils due to their presence in soil minerals that then accumulate as a result of soil-forming processes. These trace elements, their content in the lithosphere, soil, and plant tissue are as follows:

| Trace Element | Lithosphere (mg/kg) | Soil (mg/kg) | Plant Tissue (mg/kg) |
|---|---|---|---|
| Aluminum (Al) | — | — | 10–200 |
| Arsenic (As) | 1.5 | 5.0 | 0.01–1.0 |
| Barium (Ba) | 500.0 | 100.0 | 10–100 |
| Cadmium (Cd) | 0.11 | <1.0 | 0.05–0.20 |
| Chromium (Cr) | 100.0 | <100.0 | 0.1–0.5 |
| Cobalt (Co) | 20.0 | 10.0 | 0.01–0.30 |
| Lead (Pb) | 14.0 | 2–200 | 0.1–5.0 |
| Mercury (Hg) | 0.05 | 0.03 | 0.001–0.01 |
| Nickel (Ni) | ~80 | 30–40 | 0.1–1.0 |
| Selenium (Se) | 0.05 | 0.1–2.0 | 0.05–2.0 |
| Silicon (Si) | 277,000 | — | — |
| Titanium (Ti) | 5,600 | 1,800–3,600 | — |
| Vanadium (V) | 160.0 | 100–1,000 | 0.1–1.0 |

Some of these trace elements are considered essential for animals whose criteria for essentiality are less restrictive than that for plants (see page 18). Therefore, their presence in plants when consumed by animals will impact their health.

Based on published data, Markert (1994) defined what he called the reference plant level for those elements found in trace amounts found in the dry weight of plant tissue.

| Trace Element | mg/kg | Trace Element | mg/kg |
|---|---|---|---|
| Aluminum (Al) | 80 | Iodine (I) | 3.0 |
| Antimony (Sb) | 0.1 | Iron (Fe) | 150 |
| Arsenic (As) | 0.1 | Lead (Pb) | 1.0 |
| Barium (Ba) | 40 | Mercury (Hg) | 0.1 |
| Beryllium (Be) | 0.001 | Nickel (Ni) | 1.5 |
| Bismuth (Bi) | 0.01 | Selenium (Se) | 0.02 |
| Bromine (Br) | 4.0 | Silver (Ag) | 0.2 |
| Cadmium (Cd) | 0.05 | Strontium (Sr) | 50 |
| Cerium (Cs) | 0.2 | Thallium (Tl) | 0.05 |
| Chromium (Cr) | 1.5 | Tin (Sn) | 0.2 |
| Cobalt (Co) | 0.2 | Titanium (Ti) | 5.0 |
| Fluorine (F) | 2.0 | Tungsten (W) | 0.2 |
| Gallium (Ga) | 0.1 | Vanadium (V) | 0.5 |
| Gold (Au) | 0.001 | | |

That found in the plant does not specify whether the elements exist in the entire plant or in all parts of the plant or in one of its parts (stems, petioles, leaves), or as to the maturity of the plant. It is well known that certain elements accumulate in conductive tissues, such as stems and petioles (nitrate N and K are good examples), in leaf margins (B and Mn are good examples), as well as tend to either decline (a good example being Al) or increase in concentration with maturity (Ca and Mg are good examples). Some of these changes reflect changing absorptive characteristics of the plant's root, or a change in the dry matter content of the plant that occurs with maturity, either diluting or concentrating a particular element.

## 15.3   HIGH SOIL CONTENT ELEMENTS

Three elements in this category—Al, Si, and Ti—can exist in plant tissues at fairly high concentrations (at percentage levels in dry matter) depending on their

- Concentration in the soil, being major constituents in most soils
- Level of availability depending on the soil pH, organic matter and P contents, and cation exchange capacity
- Their ease of absorption and translocation associated with particular plant species

Of these three elements, Al is the only element that may be detrimental to plant growth when existing at a high concentration in the soil solution, that is then root absorbed and accumulates in plant tissues in high concentrations (over 200 ppm for some plants would suggest toxicity). Aluminum availability is high in acid soils (water pH <5.4), and low in soils high in organic matter and P contents. Interestingly, Al is readily absorbed by plant roots following seed germination and during early plant growth; but as the plant matures, absorption is restricted and the Al content in plant tissues significantly declines. If plant roots are physically damaged, damaged by disease or insects, or subjected to periods of anaerobic conditions, Al is easily absorbed, and when translocated within the plant, its presence in plant tissues can jeopardize plant health. For Al as well as for other trace elements, after absorption, translocation within the plant may be restricted, with high concentrations found in the roots, but not in stems, petioles, leaves, or fruit.

## 15.4   AVAILABILITY FACTORS

The factors influencing plant root absorption of trace elements are no different from that associated with the micronutrients. The combined effect of soil pH, content of interacting elements, texture, and organic matter content affects their availability, and therefore the potential for root absorption. Within the soil itself, the processes of mass flow, diffusion, and root interception also relate to their availability for root absorption, as it does for the major elements and micronutrients (see page 12).

## 15.5   ACCUMULATOR PLANTS AND ELEMENTS

There are plants known as "accumulator plants" that are able to absorb sizeable quantities of an element from the rooting media. Three trace elements and plant species that fit the accumulator category are

- Cadmium (Cd): lettuce, spinach, celery, cabbage, sunflower
- Lead (Pb): kale, ryegrass, celery
- Nickel (Ni): sugar beets, ryegrass, marigold, onion

## 15.6   SYMBIOTIC ELEMENT

Cobalt (Co) is an element that is required by organisms that are able to fix atmospheric nitrogen ($N_2$). Therefore, some would consider Co as an essential element for those legumes that are able to symbiotically fix $N_2$ that exists in the atmosphere surrounding the plant roots.

# Section IV

## Methods of Soil Fertility and Plant Nutrition Assessment

# 16 Soil Testing

"There is good evidence that the competent use of soil tests can make a valuable contribution to the more intelligent management of the soil,"—a statement that was included in the National Soil Test Workshop Report (Nelson et al., 1951). The development and use of soil testing procedures as a means for determining lime and fertilizer amendments required to improve and sustain a fertile soil began in the early 1940s. Fitts and Nelson (1956) wrote about the history of soil testing from those initial times up to 1956. The book by Jones (2001) and the *Handbook on Reference Methods for Soil Analysis* (Anonymous, 1992) include historical information on soil test development and its use for assessing the fertility status of soils.

## 16.1 PURPOSES

The purposes for a soil test suggested by Fitts and Nelson (1956) are to

1. Group soils into classes for the purpose of suggesting fertility and lime practices
2. Predict the probability of getting a profitable response in the application of plant nutrient elements
3. Help evaluate soil productivity
4. Determine specific soil conditions that may be improved by the addition of soil amendments or cultural practices

The first three purposes are widely practiced, while the fourth is little understood and rarely used. In addition, using a sequence of soil test results to "monitor" the changing soil fertility status of a cropped soil is rarely used (see page 125).

A soil test is conducted in a sequence of steps: sampling, sample preparation, laboratory analysis, analytical result, and interpretation of results—as is illustrated in Figure 16.1.

## 16.2 FIELD SAMPLING

A soil test result is no better than the techniques and care used to collect the soil sample and to prepare it for laboratory analysis. There are three techniques for collecting a soil sample

1. Simple random sampling
2. Stratified random sampling: selecting individual soil cores in a random pattern within a designated area
3. Systematic or grid sampling

Which of these sampling procedures is used will depend on the use for the determined soil test result.

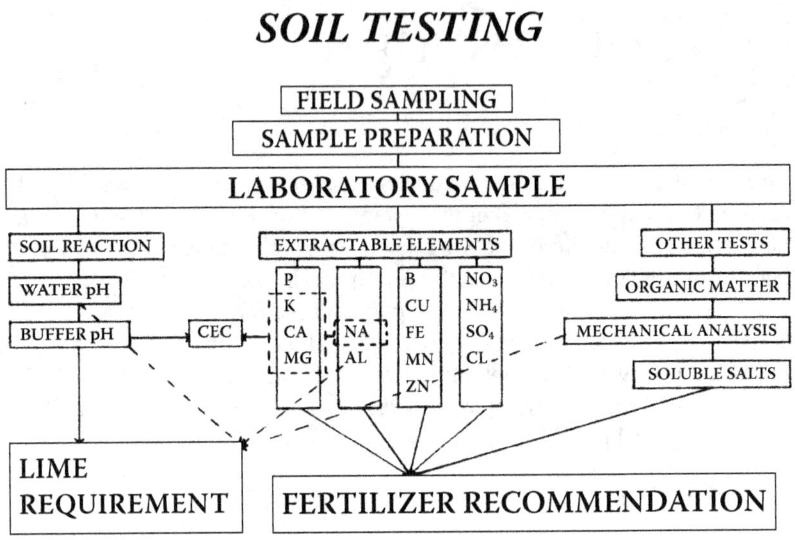

# *SOIL TESTING*

| FIELD SAMPLING |
| SAMPLE PREPARATION |

## LABORATORY SAMPLE

| SOIL REACTION | | EXTRACTABLE ELEMENTS | | | | OTHER TESTS |

| WATER pH | | P | | B | NO₃ | | ORGANIC MATTER |

| BUFFER pH | CEC | K | | CU | NH₄ | | MECHANICAL ANALYSIS |
| | | CA | NA | FE | SO₄ | |
| | | MG | AL | MN | CL | SOLUBLE SALTS |
| | | | | ZN | | |

| LIME REQUIREMENT | FERTILIZER RECOMMENDATION |

**FIGURE 16.1**   Sequence of procedures for conducting a soil test.

If the goal is to determine the soil test level for a cropped soil in order to formulate a lime and fertilizer recommendation, then random sampling is the procedure to follow. Systematic or grid sampling is recommended when precision agricultural management procedures are being used, allowing for adjustment to the rates of lime and fertilizer applications based on the fertility status by designated areas.

The number of sample cores per area is not a fixed number or designated area, although the recommendation of fifteen to twenty cores taken at random from a 100-acre area is a common recommendation, with more detailed sampling depending on the homogeneity of the area the soil sample is to represent. However, the mean value of a multiple of test results gives a better evaluation of the fertility status of a soil than can be obtained from a single sample result.

Avoidance procedures are equally applicable: not mixing soil types, not sampling from either low or high spots in the field, along fence, tree, or roadways, or in those areas not typical of the total area. These avoidance criteria are less important if the area is being grid sampled.

Coring devices are recommended for collecting soil samples, including a bucket auger, Brown soil probe, Oakfield auger, or Hofer probe, although the use of a tilling spade or similar tool can be used.

Coring should be just to the plow depth or depth of the surface soil, or the expected rooting depth where 75% of the plant roots are to be found (for pastures, turf, and orchards).

## 16.2.1 BEST TIME TO SOIL SAMPLE

Because soil is an ever-changing dynamic substance and its chemical characteristics are influenced by both moisture and temperature as well as cropping intensity, it is best to collect soil samples at a set time each year. There is probably no optimum time, but for consistent results that best represent what the crop "sees," soil sampling during the crop growing season, preferably at the time the crop begins to flower or set fruit, is recommended. Following such a sampling routine works best when soil test results are being tracked over time (see page 125). Sampling prior to establishing a crop is recommended when previous soil test information is not available

### 16.2.2 SUBSOIL SAMPLING

In general, the sampling depth is confined primarily to what is termed the "surface soil," probing to that depth which is the mixing depth when tilling the soil, or to that depth at which the majority of crop roots exist. However, the subsoil can be a significant source of water and essential plant nutrient elements that can contribute to higher crop yields.

With the tillage equipment available today, both lime and fertilizer can be applied to considerable depths, making for a deep homogenous rooting profile. Therefore, the deeper uniform conditions exist in the soil, the more productive the soil can be, allowing for greater depth penetration of plant roots. The objective should be to make the soil as uniform in pH and plant nutrient elements as possible based on subsoil physical and chemical conditions and the equipment available for deep mixing. Tillage that brings a portion of the subsoil into the surface soil may significantly change the fertility status of the surface soil, therefore requiring additional treatment to establish uniformity.

Essential plant nutrient elements can leach into the subsoil with time if the surface soil is kept in a state of high fertility. When plant roots do venture into the subsoil, they can "mine" those elements that exist, leaving the subsoil more acidic and lower in essential element status, unless means are taken to replenish the subsoil by deep tillage liming and fertilization.

### 16.2.3 SOIL PREPARATION FOR LABORATORY SUBMISSION

If the collected soil cores are wet, they should be allowed to partially air-dry before placing them in the soil sample box or bag used to transport the sample to the laboratory. Collected soil cores should not be oven dried. The sample box or bag should allow for easy air exchange, and therefore not be placed in a container that would create an anaerobic condition in the collected cores.

## 16.3 SOIL LABORATORY SELECTION

It is important that the soil testing laboratory selected to conduct the soil test be using those procedures that conform to the physical and chemical characteristics of the soil received. A common error is for the farmer/grower to seek out a soil testing laboratory based on cost of services and/or stature, rather than ensuring that the testing procedures used conform to the soil type to be tested. Using a testing procedure not applicable will give results that can be misleading in interpretation and recommendations.

Most soil testing laboratories offer an array of testing procedures, such as a basic or standard test that includes soil water pH and level (availability) of the major essential elements, P, K, Ca, and Mg. Tests for the micronutrients B, Cu, Fe, Mn, and Zn may be offered, either individually or as a group, again depending on the soil type and extraction procedures suitable for such determinations. A micronutrient assay result can be misleading if the soil type does not conform to the physiochemical properties (texture, organic matter content, pH, soil solution properties) associated with its potential insufficiency as well as crop sensitivity (see Chapter 12, "Micronutrients Considered Essential to Plants"). Most micronutrient soil tests have associated factors required for proper interpretation, such as pH, organic matter content, and soil texture, soil factors that must be factored in when evaluating a micronutrient test result.

For acid soils, the buffer pH is usually given along with a lime requirement recommendation. Some laboratories will calculate the cation exchange capacity (CEC) based on a summation of the cations, or on an actual CEC determination procedure. Determinations of organic matter content and texture may also be additionally offered tests.

Some soil testing laboratories have developed their own soil testing procedures that conform to certain soils that have unique physical and/or chemical properties, procedures justified on the basis of research that correlates the obtained test value with crop response.

Comparing soil test results obtained from different laboratories can also be confusing if one is not familiar with soil test methodology. Some will depend more on soil test results obtained from a local laboratory or services offered by the state than those generated by a commercial soil testing laboratory in the area, or some distance away. It is best to select a laboratory and continue to use its services if assay results are to be tracked over time. Most soil testing laboratories are competent in their ability to assay the soil sample submitted. Standardization of soil test methodology has occurred and most laboratories participate in verification programs based on periodically submitted soil samples.

## 16.4 LABORATORY SOIL TESTING PROCEDURES

Those procedures associated with soil pH and lime recommendation methodology are discussed in Chapter 9, "Soil pH: Determination and Interpretation." Those soil extraction reagents and procedures for determining the elemental status of a soil are given in Appendix B.

## 16.5 INTERPRETATION OF A SOIL TEST RESULT

A soil test result either serves as a means of defining the fertility status of the soil or the relationship that exists between the soil test value and crop growth, or that may be expressed as a probable response to an application of that element to the soil in the form of fertilizer. There is no established system for making these evaluations. For most micronutrients, there is no simple relationship between test level and expected crop response without considering soil factors such as pH, texture, and organic matter content, as well as plant requirements. Various methods have evolved for interpreting a soil test result; some have been standardized, while others have not. Some laboratories have developed their own system for interpreting a soil result using various systems of either word designation, percentage expected response to fertilizer applications of that element, or various ratio systems primarily associated with the major cations K, Ca, and Mg.

## 16.5.1  Word Designation

There has developed a word designation system that is correlated to a predicted crop response as given in the following table:

**Predicted Response of a Crop to the Amount of Extracted Plant Nutrient Element by a Soil Test**

| Soil Test Level | Predicted Crop Responses/Fertilizer Requirement |
|---|---|
| Very high | No crop responses predicted for fertilization with that particular element. Crop may be adversely affected by addition of this element as fertilizer. |
| High | No crop response predicted for fertilization with this particular element. May apply that element sufficient to just meet the crop requirement. |
| Medium | 75% to 100% of maximum expected yield predicted without fertilization. Low to medium amounts of fertilizer may be needed for economic maximum yield and/or quality for high-value crops. |
| Low | 50% to 75% of maximum expected yield predicted without fertilization. Needs modest to high fertilizer rates to ensure crop nutrient element sufficiency for the element. |
| Very low | 25% to 50% of maximum expected yield predicted without fertilization. Without adding this element, deficiency symptoms may occur. High rates of fertilizer are required. |

When the soil test level is less than medium, following a generated fertilizer recommendation may not result in an anticipated yield response due to factors associated with the fertilizer itself (see Chapter 20, "Organic Fertilizers and Their Properties"), and time and method of application (see Chapter 22, "Soil Water, Irrigation, and Water Quality").

There are other ways of interpreting a soil test result based on crop response as well as having environmental implications, such as outlined below:

**Soil Test Categories and Recommendations Based on Crop Response and Environmental Impact**

| Category Name | Category Definitions | Recommendations |
|---|---|---|
| | **Crop Response** | |
| Below Optimum | Element considered deficient | Based on crop requirement will build soil fertility into optimum soil fertility into optimum range over time |
| (very low, low, medium) | Will probably limit crop yield | Starter fertilizer recommended |
| | High to moderate probability of economic response | No additional fertilizer for current crop needed |

| Optimum | Element considered adequate | Fertilizer often recommended to maintain soil in the optimum test range |
| (sufficient, adequate) | Will probably not limit yield | Starter fertilizer may be recommended for some crops |
| | Low probability of economic yield response to added fertilizer | |
| Above optimum | Element more than adequate | No additional fertilizer recommended |
| (high, very high) | Will not limit yield | Remedial action may be needed to prevent phytotoxicity or environmental problems |
| | Very low probability of economic yield response | |
| | Very high levels, probability of negative impact on crop fertilizer added | |

**Environmental Response**

| Potential negative environmental impact | Higher potential to cause environmental degradation | Minimize environmental impact |
| (very high, excessive) | Should be monitored closely | Some additions may be needed for crop response |
| | Liklihood of environmental problems depends on site-specific characteristics | Fertilizer additions, including starter fertilizer, not recommended crop response guidelines |
| | May be above or below optimum level based on crop response | Remedial action may be required to protect the environment |

## 16.5.2 CRITICAL VALUES

Another method of soil test interpretation is the use of critical values, an interpretation method that is mostly limited to P and several of the micronutrients, and only for those micronutrients when the soil type factors are factored into the interpretation. For example, critical values for P have been established based on soil test methodology:

| | Critical Value (lbs P/A) | |
| Soil Test Method | Deficient | Excess |
| --- | --- | --- |
| Morgan | <3.5 | >6.5 |
| Bray P1[a] | <30 | >60 |
| Bray P2[a] | <30 | >60 |
| Olsen | <11 | >22 |
| AB-DTPA | <15 | >30 |
| Mehlich No. 1 | <30 | >100 |
| Mehlich No. 3 | <36 | >90 |

[a] Frequently, both Bray P1 and Bray P2 tests are performed and the test results are compared.

### 16.5.3 Ratio Concept of Soil Interpretation

A ratio concept, frequently referred to as base saturation, assesses the fertility status of a soil based on what portion of the cation exchange capacity (CEC) is occupied by each of the major cations, $K^+$, $Ca^{2+}$, and $Mg^{2+}$. The concept has relevance primarily for soils having cation exchange capacities greater than 10 meq/100g.

The desired percentage ranges of the three cations on the cation exchange complex are

- Calcium (Ca) – 60% to 80%
- Magnesium (Mg) – 15% to 20%
- Potassium (K) – 5% to 10%

This concept was once widely accepted, but has been found wanting since attempting to adjust the cation ratio is not easily done. Although the ratio among the cations in the soil solution does affect the root absorption of any one of the cations, it is not the primary factor affecting availability, as is discussed in Chapters 2, 4 and 8. In addition, the cation ratio concept applies primarily to soils that fall within a narrow soil cation exchange capacity range.

There are ratio concepts that have proven useful when dealing with particular soils and crops. For example, the Hartz ratio, the amount of exchangeable soil K divided by the square root of soil Mg in meq/100g units $[K^+/Mg^{2+})^{1/2}]$, can be used to predict the potential for the occurrence of tomato fruit color disorders, high occurrence likely when the ratio is less than 0.30. This ratio has wide application for the production of both processing and fresh market fruit being grown in the lakebed soils of Indiana, Ohio, and Michigan as well as certain soil types in California and Pennsylvania. A Hartz Ratio calculator can be found at http://www.oardc.ohio-state.edu/tomato/HartzRatioCalculator.htm.

## 16.6 SOIL TEST RESULT TRACKING (MONITORING)

By graphing soil test results versus time, the track can be helpful in assessing the effect that liming, fertilizing and cropping procedures are having on the fertility status of the soil. Some will use this information to evaluate the effect of past treatment and cropping sequences on soil test levels, and then establish management procedures to obtain a particular soil fertility level (pH and level of the essential plant nutrient elements), and develop a strategy to maintain that established soil fertility level with fertilizer additions being made based on crop requirements.

## 16.7 LIMING AND FERTILIZER USE STRATEGIES

Lime and fertilizer recommendations are frequently based on two concepts:

1. Adding sufficient amounts to meet the crop requirement with additional amounts added to bring the soil test level for an element to a higher test level
2. Adding sufficient amounts to meet the crop requirement as well sufficient amounts to maintain the current soil test level

There is actually a third strategy when the current soil test level is high or excessive, excluding the element from the fertilizer recommendation.

A fertilizer recommendation may be modified based on the method of application (see Chapter 21, "Fertilizer Placement") or the time of application.

Nitrogen fertilizer recommendations are usually based on crop requirement, some making a modification based on the previous crop and/or the organic matter content of the soil.

# 17 Plant Analysis and Tissue Testing

Plant analysis, sometimes referred to as either "leaf analysis" or even "tissue testing," although actually a different methodology (see below), is a technique that determines the elemental content of a particular plant part taken at a specific time during the growth of a plant. To avoid the possibility of confusion in terminology, the term "plant analysis" will be used, although the actual sampling procedures involve the assay of either collected leaves or petioles rather than the plant itself.

Plant analysis is used as a means of determining and evaluating the essential plant nutrient element status of the plant, a procedure that involves a series of steps as illustrated in Figure 17.1.

## 17.1 PLANT ANALYSIS OBJECTIVES

Plant analysis objectives include

- Verification of an elemental insufficiency (deficient or excess)
- Evaluation of the elemental status of the soil/crop environment
- Determination of essential element needs for the sampled plant
- Crop logging to monitor changes in elemental status of a plant with time
- Determining the content of nonessential elements in the plant that may affect its growth or utilization
- As a basis for a fertilizer recommendation for that crop or future crops

## 17.2 SEQUENCE OF PROCEDURES

A plant analysis is carried out in a series of four steps:

1. Sampling
2. Sample preparation
3. Laboratory analysis
4. Interpretation

Each step in the sequence is equally important to the success of the technique. A diagram of this sequence is given in Figure 17.1.

127

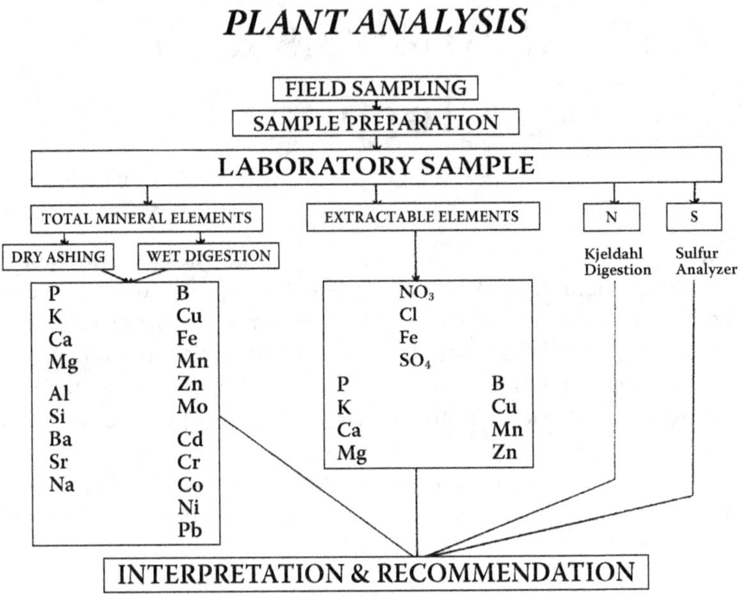

# PLANT ANALYSIS

| FIELD SAMPLING |
| SAMPLE PREPARATION |

## LABORATORY SAMPLE

| TOTAL MINERAL ELEMENTS | EXTRACTABLE ELEMENTS | N | S |

| DRY ASHING | WET DIGESTION |    Kjeldahl   Sulfur
Digestion  Analyzer

| P | B |
| K | Cu |
| Ca | Fe |
| Mg | Mn |
| Al | Zn |
| Si | Mo |
| Ba | Cd |
| Sr | Cr |
| Na | Co |
| | Ni |
| | Pb |

NO₃
Cl
Fe
SO₄

| P | B |
| K | Cu |
| Ca | Mn |
| Mg | Zn |

| INTERPRETATION & RECOMMENDATION |

**FIGURE 17.1**    Sequence of procedures for conducting a soil test.

## 17.3   SAMPLING TECHNIQUES

Plant species, plant age, plant part, and time sampled are variables that affect the interpretation of a plant analysis result, and therefore the need for careful sampling. In addition, most of the essential elements are not equally distributed in the plant or within its parts. Sampling instructions are quite specific in terms of plant part and stage of growth, as a comparison of an assay result with established critical, standard values or sufficiency ranges are based on a clearly identified plant part taken at a specific time. It should be emphasized that an analysis of a different plant part or a plant part taken at a time other than that specified may not be easy to interpret using established interpretative data. When no specific sampling instructions are given or are unknown, the general rule of thumb is to select upper mature leaves. Procedures for collecting a plant tissue sample for either field or laboratory determination of its elemental content have been widely published (Mills and Jones, 1996; Jones, 2001).

The important components for proper plant tissue collection are:

- Definite plant part taken at a specific location on the plant
- Stage of plant growth or specific time of sampling
- Number of parts taken per plant
- Number of plants selected for sampling

Following the sampling directions prescribed, the sampler should achieve reasonable statistical reliability.

## 17.3.1 WHEN TO SAMPLE

When to take a plant tissue sample varies with plant species and the objective of the plant analysis. For evaluating the elemental status of a plant, taking a tissue sample during the vegetative period is normally advised, selecting recently matured tissue. Sampling may also be recommended when the plant is beginning its reproductive stage, just before setting flowers and/or fruit. When to sample is determined by what interpretative data is available that will have a sampling time designation.

## 17.3.2 NUMBER OF SAMPLES AND PLANTS TO SAMPLE

Precision requirements will dictate the number of plant parts to be collected as well as the number of plants to be sampled in order to make a reliable composite sample, or just how many composite samples will be necessary to ensure sufficient replication. Various studies indicate that the number of individual leaves and/or plants required is correlated with the desired variance to be obtained. The combination of the number of plants selected for sampling plus the number of samples per plant determines the variance associated with the final analysis result. The variance is more significantly affected by the number of plants sampled rather than by the number of leaves collect per plant. Normally, the mean value of several composite sample assays is a more accurate estimate than a single assay result based on a single composite sample consisting of the same total number of individual samples.

## 17.3.3 LACK OF HOMOGENEITY

How a plant tissue sample is selected is important because the distribution of the essential plant nutrient elements within the plant, and even within any one of its parts, is not homogeneous due to a number of factors. For example, as plant tissues mature, there are changes due to

- The movement of mobile elements from the older tissue to newly developing tissues
- An accumulation of nonmobile elements
- A reduction in dry matter content

One sign of increasing maturity in leaves is an increasing concentration (accumulation) of Ca and Mg and a decreasing concentration (reduction) of N, P, and K.

Another factor contributing to this variation is the relative proportion of leaf blade to mid-rib and the size of the leaf, anatomical factors that can affect the concentration of elements found in the whole leaf. For example, the leaf mid-rib will normally contain a higher concentration of K than the blade. Similarly, the relative proportion of leaf blade to margin affects the B and Mn contents of the whole leaf because these two elements accumulate to fairly high concentrations in the leaf margin.

Compound leaves pose a problem for sampling because leaflets mature at different times. For tomato, some recommend that the whole leaf be sampled, while the author has found that the end leaflet is a better indicator of elemental status, and less likely to give a false estimate of the elemental status of the whole plant.

A sampling procedure that would enhance the distribution effect of elements within the leaf will affect the analysis result. For instance, a sampling procedure in which only the leaf tips are sampled or blade punches are collected will produce a different analytical result compared to the assay of the entire intact leaf.

### 17.3.4 PETIOLES

Petioles are not part of the leaf blade and should not be included if the leaf is the desired sample. For some crops, such as grape and sugar beet, the petiole may be the plant part to be assayed rather than the leaf blade. Petioles, being conductive tissue, are higher in certain elements, such as K, P, and nitrate nitrogen ($NO_3$-N), than the attached leaf blade.

### 17.3.5 COMPARATIVE PLANT TISSUE SAMPLES

Sampling two different populations of plants for comparative purposes can also be difficult, particularly when some type of stress has resulted in substantial differences in growth characteristics. When two or more sets of plants exhibit varying signs of a possible elemental insufficiency, collecting tissue for comparative purposes from affected and unaffected plants is desirable, but it can be difficult due in part to the effect that elemental stress had on plant growth and development. It is important, whenever possible, to obtain plant tissue samples when the symptoms of stress first appear. Therefore, great care must be taken to ensure that representative samples are collected for such comparisons, and that the interpretation of the analysis result takes into consideration the condition of the plants when the tissue samples were collected, whether normal in physical appearance or not, due to some type of stress.

### 17.3.6 WHAT NOT TO SAMPLE

Avoidance criteria are also crucial. Plants to be avoided are ones that

- Have been under long climatic or nutritional stress
- Have been damaged mechanically or by insects
- Are infested with disease
- Are covered with dust or soil, or foliar-applied spray materials unless these extraneous substances can be removed effectively (decontamination procedures are discussed in 17.4.3)
- Are border row plants or shaded leaves within the plant canopy
- Contain dead plant tissue
- Are whole plants unless specified for that crop plant
- Plant roots unless they are selected for a specific purpose from hydroponically grown plants

Although it is possible to assay just about any plant part, or even the whole plant itself, the biological significance of an analysis result depends on the purpose of the assay, and on the availability of interpretative data for the plant part collected and assayed. For example, the assay of fruits and grain, or an analysis of the whole plant or one of its parts at maturity or at harvest, does not usually provide reliable information on the elemental status of the plant during its growth period.

When conducting a plant analysis, the primary objective should be to obtain that plant part for which assay results can be compared to known interpretative values, or plant tissue collected from plants exhibiting symptoms and those without symptoms at the same stage of growth, basing the interpretation on comparing the assay results.

### 17.3.7   COLLECTING A SOIL SAMPLE

A soil sample or samples should be taken at the time and in the same vicinity where plant tissue samples are collected, and then the soil assay results included with the plant analysis data. Such soil test information can be particularly helpful to the interpreter of a plant analysis result. Recent soil test data may be sufficient, and therefore it may not be necessary to collect a soil sample. However, when a plant analysis is collected for diagnostic purposes, a soil sample taken at the same time and in the same area can provide valuable information when evaluating the plant analysis result.

## 17.4   PLANT TISSUE HANDLING, PREPARATION, AND ANALYSIS

Once a plant sample is collected, care is required to ensure that the tissue retains its original condition, preventing loss of dry weight and keeping the tissue from being contaminated by contact with sampling tools and storage containers. Placing fresh plant tissue in an enclosed container for any length of time requires refrigeration, with the sample container being free from any substances that can contaminate the sample.

### 17.4.1   DRY WEIGHT PRESERVATION

It should be remembered that an elemental concentration interpretation is based on tissue dry weight; therefore, any condition that affects dry weight will in turn affect the elemental concentration. If collected plant tissue begins to decay, a significant reduction in dry weight will occur as well as the loss by volatilization of some elements, particularly N and S. In order to preserve tissue dry weight,

- Do not place fresh plant tissue in plastic bags unless the temperature is kept below 40°F (5°C).
- Air drying tissue prior to shipment to the plant analysis laboratory will minimize the loss in dry weight.
- If possible, deliver the plant tissue to the laboratory within 24 hours of collection.

## 17.4.2   Sources of Contamination

Soil, dust, and foliar-applied chemicals are common sources of contamination, and if present and not removed, can result in a misinterpretation of an analytical result. Contamination can occur as the result of careless handing of a collected tissue sample. Those elements found as major constituents in soil and dust will alter the analytical result for those elements, the elements being Al, Fe, and Si. If all three of these elements appear in high concentration in the assay result, soil/dust contamination is likely their source; but if only one element appears at high concentration, soil/duct contamination is not likely the source. Under some circumstances, assay results for other elements may be affected, depending on the nature and degree of presence of the soil/dust.

Some fungicides and other pest control chemicals may contain elements, that will appear in the assay result. Copper, Mn, and Zn are common ingredients in such pest control chemicals. Therefore, when making an interpretation of a plant analysis result, the interpreter needs to know if such chemicals have been foliar applied.

## 17.4.3   Decontamination

Decontamination to remove foreign substances from the surface of tissue may be necessary, particularly if they contain elements that are essential for the interpretation of the analysis result. Normally, decontamination is

- Required when plant tissue is covered with soil and dust particles
- When coated with foliar-applied materials that contain elements of interest in the plant analysis determination

If Fe is an element of primary interest, the plant tissue must be decontaminated; otherwise, the Fe analysis result will be highly suspect. If Fe is not an element of primary interest, rainfall or the application of overhead irrigation will normally keep just-maturing leaf tissue relatively free from a significant accumulation of dust and soil particles; therefore, decontamination may not be necessary. However, if plants are not regularly bathed with either rainfall or overhead irrigation, dust and soil particles will begin to accumulate on leaf tissue, requiring decontamination in order to remove the accumulated particles.

The characteristics of the tissue surface will determine the effectiveness of the decontamination process. Rough-surfaced tissue or tissue that has a pubescent surface may be impossible to decontaminate. In some instances, the decontamination procedure itself may significantly alter the elemental composition of the tissue. The procedure may add elements to the tissue, or it may leach elements such as K and B.

The following procedure is recommended for decontaminating a plant tissue sample:

- Use only fresh, fully turgid plant tissue.
- Wash—quickly (20 to 30 seconds)—fresh, fully turgid tissue in a mild 2% phosphorus-free detergent solution, one leaf at a time.

- Rub the tissue surface gently with the fingers.
- Quickly wash the detergent solution from the tissue surface in flowing pure water.
- Shake to remove the excess water and then oven-dry at 176°F (80°C).

### 17.4.5   ELEMENTAL ANALYSIS PROCEDURES

This subject is discussed in Appendix C.

### 17.4.6   ELEMENTAL CONTENT

Average concentrations for the essential plant nutrient elements in plant dry matter sufficient for adequate plant growth relative to the micronutrient Mo were given by Epstein (1965):

| Element | mmol/g | mg/kg (ppm) | % | Relative No. of Atoms |
|---|---|---|---|---|
| Molybdenum (Mo) | 0.001 | 0.1 | — | 1 |
| Copper (Cu) | 0.10 | 6 | — | 100 |
| Zinc (Zn) | 0.30 | 20 | — | 300 |
| Manganese (Mn) | 1.0 | 50 | — | 1,000 |
| Iron (Fe) | 2.0 | 100 | — | 2.000 |
| Boron (B) | 2.0 | 20 | — | 2,000 |
| Chlorine (Cl) | 3.0 | 100 | — | 3,000 |
| Magnesium (Mg) | 80 | — | 0.2 | 80,000 |
| Phosphorus (P) | 60 | — | 0.2 | 60,000 |
| Calcium (Ca) | 125 | — | 0.5 | 125,000 |
| Potassium (K) | 250 | — | 1.0 | 250,000 |
| Nitrogen (N) | 1,000 | — | 1.5 | 1,000,000 |

### 17.4.7   EXPRESSION OF ANALYTICAL RESULTS

Normally, a plant analysis result is expressed as a percent (%) of the dry weight of tissue for the major elements (N, P, K, Ca, Mg, and S), and in parts per million (ppm) or milligrams per kilogram (mg/kg) for the micronutrients (B, Cu, Fe, Mn, Mo, and Zn).

Some express a plant analysis result in either milliequivalents (m.e.) or microequivalents (p.e./100g). To convert percent (%) and parts per million (ppm) to these other values, the following are the conversion factors:

| Element | Converting from | Valance | Equivalent Weight | Factor |
|---|---|---|---|---|
| Nitrogen (N) | % to m.e. | 3 | 4.6693 | 214.6 |
| Phosphorus (P) | % to m.e. | 5 | 6.1960 | 161.39 |
| Potassium (K) | % to m.e. | 1 | 39.096 | 25.578 |
| Calcium (Ca) | % to m.e. | 2 | 20.040 | 49.508 |
| Magnesium (Mg) | % to m.e. | 2 | 12.160 | 82.237 |

| Boron (B) | ppm to p.e. | 3 | 3.6067 | 27.726 |
| Copper (Cu) | ppm to p.e. | 2 | 31.770 | 3.1476 |
| Iron (Fe) | ppm to p.e. | 3 | 18.617 | 5.3726 |
| Manganese (Mn) | ppm to p.e. | 2 | 27.465 | 3.6410 |
| Zinc (Zn) | ppm to p.e. | 2 | 32.690 | 3.0590 |
| Sulfur (S) | % to m.e. | 2 | 16.033 | 62.577 |
| Sodium (Na) | % to m.e. | 1 | 22.991 | 43.496 |
| Chlorine (Cl) | % to m.e. | 1 | 35.457 | 28.171 |

*Note:* Millequivalents can be converted to percentages by multiplying by the equivalent weight/1,000, and microequivalents can be converted to parts per million (ppm) by multiplying by the equivalent weight/100. Factor × % = m.e./100 g and factor × ppm = p.e./100 g.

Four methods for expressing essential plant nutrient element concentrations in plant tissue are as follows (numbers have been selected for illustrative purposes):

| Element | Dry Weight Basis | | | |
| --- | --- | --- | --- | --- |
| | Percent (%) | g/kg | cmol(p+)/kg | cmol/kg |
| Nitrogen (N) | 3.15 | 31.5 | 225 | 225 |
| Phosphorus (P) | 0.32 | 3.5 | — | — |
| Potassium (K) | 1.95 | 19.5 | 50 | 50 |
| Calcium (Ca) | 2.00 | 20.0 | 25 | 50 |
| Magnesium (Mg) | 0.48 | 4.8 | 10 | 20 |
| Sulfur (S) | 0.50 | 5.0 | — | — |
| | ppm | mg/kg | cmol(p+)/kg | mmol/kg |
| Boron (B) | 20 | 20 | — | 1.85 |
| Copper (Cu) | 12 | 12 | 0.09 | 1.85 |
| Iron (Fe) | 111 | 111 | 0.66 | 1.98 |
| Manganese (Mn) | 55 | 55 | 0.50 | 1.00 |
| Zinc (Zn) | 33 | 33 | 0.25 | 0.50 |

## 17.5　METHODS OF INTERPRETATION

Difficulties have been encountered in the use and interpretation of plant analyses, although the quantitative association between an absorbed essential plant nutrient element and plant growth has been widely studied. Questions raised at the *1959 Plant Analysis and Fertilizer Problems Colloquium* regarding the limitations of the plant analysis technique are still applicable today:

- The reliability of interpretative data
- The utilization of ratio and balance concepts
- Hybrid influences
- Changing physiological processes that occur at varying elemental concentrations

Today, reliable interpretative data are still lacking for

- The micronutrient Cl

- Most of the essential elements for ornamental plants
- Most plants during their very early stages of growth
- Identification of elemental concentrations considered excessive or toxic

It is also questionable whether the determination of the total Fe concentration in a particular plant tissue can be used to establish the degree of Fe sufficiency due to the problem of contamination and the characteristics of Fe within the plant itself. It has been suggested that Fe removed from the plant tissue by extraction, gives a better evaluation of the Fe status of a plant. Such consideration is mainly applicable to those plant species that are considered Fe sensitive.

An interpretation of the meaning of the elemental concentration found in the assayed plant tissue is based on a comparison with values obtained for the same type of tissue, sampling time, and plant species. The comparison is made between the laboratory analysis result and one or more of these known values or ranges in order to access the plant's elemental status. There are three sets of values that are available for making such an interpretation:

1. Critical values
2. Standard values
3. Sufficiency ranges

and each has its own characteristics.

### 17.5.1 CRITICAL VALUES

A critical value would separate sufficiency and toxicity. Therefore, one could divide critical values into two categories:

- *Lower Critical Level:* That concentration of an element in plant tissue below which deficiency occurs and above which is sufficiency
- *Upper Critical Level:* That concentration of an element in plant tissue below which sufficiency occurs and above which excess and/or toxicity occurs

This single value is difficult to use when a plant analysis result is considerably above or below the critical value. Some have suggested that the twin transition zone be used to designate the range in elemental concentration that exists between deficiency and sufficiency. Others have termed this range in concentration as the critical nutrient range (CNR). This concentration range lies within the transition zone, a range in concentration in which a 0% to 10% reduction in yield and/or plant growth occurs, with the 10% reduction in yield specifying the critical value of the element. Critical values were obtained from the assay of plant tissue at that point when visual stress symptoms appeared that could be related to a plant nutrient element insufficiency. Several concentration critical values have wide application, such as 15 ppm Zn, a reliable critical concentration value between sufficiency and deficiency for many plant species.

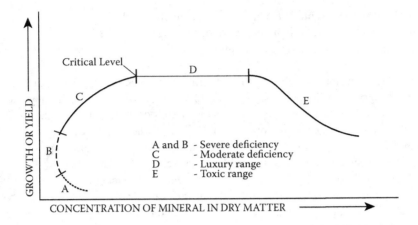

**FIGURE 17.2** General relationship between plant growth or product yield and the essential nutrient element content of a plant. (*Source:* Smith, V.R., 1962. *Annu. Rev. Plant Physiol.,* 13:81. With permission.)

Typical critical values that would have fairly wide application are:

| Critical Plant Tissue Values between Deficiency and Sufficiency[1] | | | |
|---|---|---|---|
| Element | % | Element | mg/kg |
| Nitrogen (N) | 2.50 | Boron (B) | 5 |
| Phosphorus (P) | 0.25 | Copper (Cu) | 5 |
| Potassium (K) | 1.50 | Iron (Fe) | 25 |
| Calcium (Ca) | 1.00 | Manganese (Mn) | 15 |
| Magnesium (Mg) | 0.25 | Molybdenum (Mo) | 0.25 |
| Sulfur (S) | 0.30 | Zinc (Zn) | 15 |

[1]General values that do not necessarily apply to all plant species.

## 17.5.2 STANDARD VALUES

Standard values are obtained by repeated yearly assays of plant tissue taken from plants that are normal in appearance, growing under well-managed conditions. Over time, the obtained assay results would define that range of elemental concentration related to sufficiency. Standard values best apply to fruit and nut trees, having little value for annual crop species. An assay result would be compared to the standard value with variance suggesting a plant nutrient element insufficiency. A grower could establish his own standard values that would directly reflect to his own soil-plant growing conditions.

## 17.5.3 SUFFICIENCY RANGE

Most who interpret plant analysis results for diagnostic purposes want to know that range in elemental concentration between deficiency and excess, that being what would be defined as the *sufficiency range*. The most extensive listing of *sufficiency ranges* for a wide range of crop plants can be found in the books by Mills and Jones (1996) and Reuter and Robinson (1997).

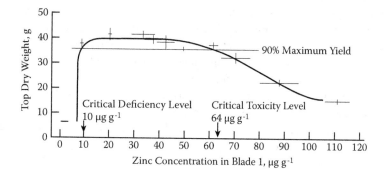

**FIGURE 17.3**   Relationship between zinc content of blade 1 of grain sorghum and top dry weight. (*Source:* Ohki, K. 1984. *Agron. J.,* 76: 253. With permission.)

Such ranges in elemental concentration have been obtained from response curves as illustrated in Figure 17.2. The C-shape of the left-hand portion of the curve has been termed the *Steenbjerg* effect, the result of the combined effect of elemental accumulation and dry matter reduction that frequently occurs when plants are under stress. If the interpreter is not familiar with this interactive relationship—element concentration and dry matter accumulation or reduction—a misinterpretation of the plant analysis may result if single concentration values are taken from such a response curve.

Sufficiency values are commonly used to interpret a plant analysis result. The *Plant Analysis Handbook II* (Mills and Jones, 1996) contains sufficiency range data for 17 categories of plants, with a total of 1,374 sets of sufficiency ranges. An example of sufficiency ranges for 5 plant species are:

| Element | \multicolumn{5}{c}{Sufficiency Ranges for Selected Crops} |
|---------|-------|----------|---------|-------|--------|
|         | Corn[a] | Soybean[b] | Tomato[c] | Apple[d] | Pecan[e] |
| | \multicolumn{5}{c}{- - - - - - - - - - - - - - - - - - % - - - - - - - - - - - - - - - - - - - -} |
| Nitrogen (N) | 2.70–4.00 | 4.00–5.00 | 2.50–5.00 | 1.90–2.70 | 1.75–3.50 |
| Phosphorus (P) | 0.25–0.50 | 0.25–0.50 | 0.35–0.50 | 0.09–0.40 | 0.10–0.30 |
| Potassium (K) | 1.70–3.00 | 1.70–2.50 | 2.50–5.00 | 1.20–2.00 | 0.65–2.50 |
| Calcium (Ca) | 0.21–1.00 | 0.35–2.00 | 1.50–3.00 | 0.80–1.60 | 0.75–1.75 |
| Magnesium (Mg) | 0.20–1.00 | 0.25–1.00 | 0.150–1.00 | 0.25–0.45 | 0.25–0.30 |
| Sulfur (S) | 0.20–0.50 | 0.20–0.40 | — | — | — |
| | \multicolumn{5}{c}{- - - - - - - - - - - - - - - mg/kg - - - - - - - - - - - - - - - - - -} |
| Boron (B) | 5–25 | 20–55 | 25–100 | 25–50 | 30–75 |
| Copper (Cu) | 6–20 | 10–30 | 5–20 | 6–25 | 10–20 |
| Iron (Fe) | 20–250 | 50–350 | 60–300 | 50–300 | 75–200 |
| Manganese (Mn) | 20–200 | 20–100 | 40–150 | 25–200 | 50–400 |
| Zinc (Zn) | 25–100 | 20–50 | 25–75 | 20–100 | 25–120 |

[a]ear leaf taken at initial silk
[b]mature leaves from new growth
[c]end leaflet from recently mature leaf
[d]mature leaves from new growth
[e]leaflet pairs from new growth

Other analysts have drawn response curves with varying slopes within the deficiency range as shown in Figure 17.3, better typefying the association between yield and a micronutrient concentration. The steep left-hand slope of such response curves poses a significant sampling and analytical problem because this indicates that a very small change in concentration results in a significant change in plant growth and/or yield. This phenomenon is primarily associated with two micronutrients (Mn and Zn), where a relatively small change in concentration—a change of 1 or 2 ppm in the leaf tissue—can define the difference between deficiency or sufficiency for these two elements.

## 17.5.4   EXPECTED ELEMENTAL CONTENT RANGE IN PLANT TISSUE

Listed below are the expected ranges in concentration for the thirteen essential plant nutrient elements in plant tissue dry matter:

| Element | Concentration Range |
|---|---|
| **Major Element** | **(%)** |
| Nitrogen (N) | 1.00–5.00 |
| Phosphorus (P) | 0.20–0.40 |
| Potassium (K) | 1.00–2.50 |
| Calcium (Ca) | 0.50–3.00 |
| Magnesium (Mg) | 0.60–1.00 |
| **Micronutrient** | **(ppm)** |
| Boron (B) | 10–50 |
| Copper (Cu) | 5–15 |
| Iron (Fe) | 70–150 |
| Manganese (Mn) | 30–100 |
| Molybdenum (Mo) | 0.1–0.25 |
| Zinc (Zn) | 20–50 |

These values represent the expected range, the specific amount found depending on the plant species, plant part, and stage of plant growth.

## 17.5.5   EXCESSIVE OR TOXIC CONCENTRATIONS

In an ever-increasing number of instances, identifying an excessive or toxic concentration level for an essential element has become as important as identifying the level considered deficient. Unfortunately, little information exists that defines that point at which sufficiency becomes toxicity.

## 17.5.6   DIAGNOSIS AND RECOMMENDATION INTEGRATED SYSTEM (DRIS)

This method uses the ratios of elemental contents to establish a series of values that will identify those elements from the most to the least deficient, an entirely different concept of plant analysis (Beverly, 1991). DRIS is based on the principle of elemental interrelationships by determining, in descending order, those elements from the most limiting to the least limiting. The survey approach, utilizing the world's published literature to plot elemental leaf concentration versus yield, is used to develop a distribution curve. In order to normalize the distribution curve, the yield component is divided into low- and high-yield groups. Investigators have suggested that the data bank for determining DRIS norms consist of at least several thousand entries and be randomly selected. They also propose that at least 10% of the population be in the high-yield subgroup. The cut-off value selected divides the low- from the high-yielding subgroups so that the high-yield data subgroup remains normally distributed for selecting the elemental concentration mean, ratio, and product of the elemental means. The ratio or product selected for calculating DRIS norms is the one with the largest variance that maximizes the diagnostic sensitivity.

The DRIS technique of interpretation is based on a comparison of calculated elemental ratio indices with established norms. The DRIS approach was designed to

- Provide a valid diagnosis irrespective of plant age or tissue origin.
- Rank nutrients in their limiting order.
- Stress the importance of nutrient balance.

Although the DRIS method of interpretation has been based primarily on the major elements, other DRIS indices have been generated that include the micronutrients B, Cu, Fe, Mn, and Zn. The system emphasizes the major elements, as the database for the major elements is considerably larger than the database for the micronutrients. Thus, a micronutrient DRIS index would be less reliable than one for a major element.

The DRIS concept of plant analysis interpretation has been compared to the sufficiency range interpretative values used in more traditional techniques. In general, most interpreters agree that both methods of interpretation have their advantages, but seem at times to work best when used together.

DRIS apparently works best at the extremes of the sufficiency range by pinpointing insufficient elements or balances of elements. It is less useful when plant nutrient levels are well within the sufficiency range. In most studies that have been designed to test the DRIS concept, users have found that DRIS is not entirely independent of either location or time of sampling, and that DRIS diagnosis can frequently be misleading and incorrect. Therefore, it is doubtful that the DRIS method of plant analysis interpretation will ever be exclusively used in lieu of the more traditional critical value or sufficiency range techniques.

## 17.6  WORD CLASSIFICATION OF ELEMENTAL CONCENTRATIONS

Word systems have been developed to associate elemental content determined by the assay of a specific plant part (leaf or petiole—place on the plant, whole top, or portion of the top) taken at a specific time (stage of growth) in the life cycle of the plant with plant growth or yield. One such word classification system uses five descriptive words—"deficient," "marginal," "adequate," "high," and "toxic"—to express the association between relative growth or production (% of maximum) and plant nutrient element concentration in the assayed plant part. The plant condition factors associated with these five words are

Deficient:
- Associated visual deficiency symptoms
- Severely reduced growth and production
- Corrective measures are required

Marginal:
- Reduction in growth and production
- Do not show visual deficiency symptoms
- Changes in fertilizer practice are required

Critical value (deficiency and toxicity):
- Optimal values at the lower and upper point of sufficiency range
- Adopt practices to keep above critical value (deficiency)
- Adopt practices to keep below critical value (toxicity)

Adequate:
- Range of concentration that does not increase or decrease growth and production
- Fertilizer practice need not change

High:
- Fertilizer should be reduced until the adequate range is obtained

Excessive or Toxic:
- Excessive concentration
- Reduced quality and growth

This word association system has not become the standard for interpreting a plant analysis result, although some system is needed that has universal acceptance. Most plant analysis laboratories have adopted their own system of interpretation based on what best suits those using their laboratory, particularly in those areas where plant analysis is a primary method for evaluating the plant nutrient elemental status of a crop and becomes the basis for formulating a soil fertility program. Good examples of this are in those areas where tree crops, such as citrus, peaches, apples, walnut, and pecans, are major crops.

## 17.7 PLANT ANALYSIS AS A DIAGNOSTIC TECHNIQUE

Plant analysis, as a diagnostic technique, has a considerable history of application. More recently, plant analysis results are being used to determine the combined soil and crop plant nutrient elemental status from which a prescribed lime and fertilizer recommendation can be made. Using a sequence of plant analyses to track or log changes in plant nutrient element status during the growing season, growers can determine when supplemental fertilizer treatments are needed. An end-of-season evaluation can be made using a plant analysis result as a means of establishing plant nutrient elemental sufficiency so that other plant-influencing factors can be examined that may have affected the final soil-plant yield/quality outcome.

## 17.8   EXPERIENCE REQUIRED

None of the interpretative procedures discussed above are infallible, and all of these interpretative techniques are best used by the most experienced individuals. The interpretation of a plant analysis is still an art in which the interpreter must use all the resources available to evaluate the nutrient element status of the plant, and then design a corrective treatment when an insufficiency is uncovered based on the data—plus some degree of intuition. In addition, being able to observe the situation in the field can be invaluable. With video and digital cameras, and computers, the interpreter can be supplied with real-time visuals that can aid considerably when an interpretation seems difficult to make based solely on the assay results.

## 17.9   DATA LOGGING/TRACKING OF PLANT ANALYSES

A useful but little-used technique for plant analysis utilization is data logging or track-ing, a technique based on a time series of analysis results. An example of seasonal tracking is the monitoring of the nitrate ($NO_3^-$) and phosphate ($PO_4^{3-}$) levels in cotton petioles, and nitrate ($NO_3^-$) and K monitoring to regulate N applications through the irrigation system. The objective for data logging or tracking is to regulate cultural practices to maintain plant elemental levels within the sufficiency range. The time plot of plant analysis results also warns of developing insufficiencies when the plotted slope of the timeline begins to move toward either boundary line of the sufficiency range. Thus, corrective measures can be taken before an insufficiency occurs that may lead to a reduction in yield and/or quality.

## 17.10   UTILIZATION OF PLANT ANALYSES FOR NUTRIENT ELEMENT MANAGEMENT

For many, a plant analysis result is viewed primarily as a diagnostic device for deter-mining which plant nutrient element in the plant tissue assay results is below or above the optimum concentration for normal plant growth. Five decades ago, four principal objectives for the utilization of a plant analysis result were put forth:

1. Aid in determining the plant nutrient element supplying power of the soil
2. Aid in determining the effect of treatment on the plant nutrient element sup-ply in the plant
3. Study of relationships between the nutrient element status of the plant and crop performance as an aid in predicting fertilizer requirements
4. Assist in laying the foundation for approaching new problems or for survey-ing unknown regions to determine where critical plant nutritional experi-mentation should be conducted

None of these objectives satisfies the primary practical field use of a plant analysis today—that is, to diagnose suspected plant nutrient element insufficiencies. However, the third objective mentioned seems to meet today's criteria for determining the fertilization needs of fruit and nut trees. In common use are programs that base fertilizer recommendations on a leaf analysis, particularly for perennials and plantation crops.

Using one or several of the systems of interpretation of a plant analysis result discussed above, the next step is to determine if an insufficiency exists and why, and then to devise a treatment (or treatments) to correct the insufficiency for the sampled crop, or to develop a strategy needed to prevent the insufficiency from occurring in a subsequent crop.

For example, if an Mg deficiency is confirmed by means of plant analysis, the interpreter must determine why it exists and what steps should be taken to correct the deficiency. The corrective action required could be affected by factors such as soil pH, level of soil-available Mg, relationship with the other major cations K and Ca, plant species sensitivity to Mg, extended weather conditions, etc. Adding fertilizer Mg, for example, would be ineffective if the soil pH is less than 5.4, or the cations in the soil are out of balance, or the weather conditions are not conducive to good plant growth.

## 17.11   TISSUE TESTING

Terminology is a problem here as the terms "tissue testing" and "plant (leaf) analysis" are frequently used synonymously. At one time, tissue testing was identified as an analysis of cellular sap extracted from either stems, stalks, or petioles, carried out in the field using chemically treated papers requiring the use of either liquid or solid reagents. Today, specific-ion devices can be used in the field to determine the concentration of K and nitrate-nitrogen ($NO_3$-N) in extracted cellular sap. There are also elemental analysis kits (see HACH.com) that can be used in the field for making the same determinations. Specific-ion devices and the HACH kits are expensive and require some skill and experience in collecting and assaying extracted cellular sap.

The challenge in using these electronic devices and field kits is not primarily associated with the chemistry involved, although they do exist and can be a source of error in the determination. Rather, it is the interpretation of the assay result along with the procedures for sampling and the condition of the tested crop plant that requires considerable skill and experience on the part of the user in order to make a proper evaluation of the obtained test result.

Today, tissue testing is primarily conducted in the laboratory, extracting an element or elements from oven-dried ground conductive plant tissue, such as stems or leaf petioles. Nitrate nitrogen ($NO_3$-N) cotton leaf petiole testing programs are available to growers where cotton is a major field crop. By taking a series of petiole samples during the vegetative stage of plant growth, the test results will advise the farmer if additional N is needed to maximize lint yield and quality. For some high-value fruit and vegetable crops, stem or petiole testing for $NO_3$-N, and sometimes for P as well as K, is used to regulate the application of N, P, or K fertilizers during the vegetative and fruiting period in order to ensure that the NPK levels in the plant are sufficient for high yield and quality product production. The laboratory procedures for making these determinations are given in Appendix C.

Interpretative values for these elements are readily available, both critical values as well as ranges in concentration that define insufficiency and sufficiency (Ludwick, 1998). These values and sufficiency ranges are specific for plant species, plant part, and stage of growth when sampled. Most of the interpretative data are for high cash crops, those crops where the plant content of N, P, and K are correlated to yield as well as the quality of produced product.

## 17.12  INDIRECT EVALUATION PROCEDURES

Plant N status can be evaluated by means of aerial infrared reflectance photography, a procedure that has its advantages, yet can also provide misleading information. Chlorophyll meters, such as the Minolta SPAD 502, are being used to access the N status of a plant, a nondestructive procedure whose value is based on a predetermined calibration between meter reading and N leaf content, a calibration that is based on leaf characteristics and stage of plant growth.

# Section V

---

*Amendments for Soil
Fertility Maintenance*

# 18 Lime and Liming Materials

Probably no aspect of soil fertility management is less understood than that associated with how best to maintain the pH (refers to what is called the *water pH*, that determined in a water slurry, see Chapter 9, "Soil pH: Its Determination and Interpretation") of an acid soil within the desired range. For an intensively cropped soil, some would advise that liming should be like fertilizing, an essential yearly requirement necessary to sustain high soil productivity by maintaining the soil pH at the desired level—that is, not allowing the soil pH to decline before applying lime. Allowing the soil pH to cycle by only periodic lime applications can lead to poor crop performance. When diagnosing poor crop performance, soil pH and past liming history should be investigated first.

## 18.1  LIMING TERMS

**Acid-forming fertilizer:** A fertilizer that is capable of lowering the pH (increasing the acidity) of the soil following application.

**Aglime:** Also known as "agstone." Calcitic or dolomitic limestone that is crushed and ground to a gradation of fineness that will allow it to neutralize soil acidity. Usually the material is ground to pass through sieves from 8- to 100-mesh range or finer.

**Agricultural liming material:** Any material that contains calcium and magnesium in forms that are capable of neutralizing (reducing) soil acidity.

**Burnt lime:** Calcium oxide ($CaO$), not found in nature but from calcium carbonate ($CaCO_3$) by heating limestone to drive off the carbon dioxide.

**Calcite:** The crystalline (having regular internal arrangement of atoms, ions or molecules characteristic of crystals) form the calcium carbonate ($CaCO_3$). Pure calcite contains 100% $CaCO_3$ (40% Ca) and is found in nature

**Calcitic limestone:** Widely used term referring to agricultural limestone with a high Ca content as calcium carbonate ($CaCO_3$) that may also contain a small quantity of Mg.

**Calcium carbonate ($CaCO_3$):** A compound that occurs in limestone, marble, chalk, and shells that can be used as a liming material.

**Calcium oxide ($CaO$):** Compound composed of Ca and O; formed from $CaCO_3$ by heating limestone to drive off $CO_2$; also known as quick lime, unsalted lime, burnt lime, caustic lime; does not occur in nature.

**Dolomite:** Limestone that contains magnesium carbonate ($MgCO_3$) approximately equivalent to the calcium carbonate ($CaCO_3$) content.

**Dolomitic limestone:** Limestone that contains more than 10%, but less than 50% dolomite, and from 50% to 90% calcite. The magnesium (Mg) in dolomitic limestone may range between 4.4% and 22.6%. Dolomitic limestone for soil acidity correction is defined by state statutes, normally expressed as the amount of $MgCO_3$ present based on its calcium carbonate equivalent (CCE).

**Effective Calcium Carbonate Equivalent (CCE):** An expression of aglime effectiveness based on the combined effect of chemical purity and fineness, a label identification that is required by some states.

**Fineness Index:** Percentage by weight of liming material particles that will Pass designated screen (sieve) sizes. A requirement that varies with state statutes in terms of percentage weight amounts that will pass through various mesh sizes.

**Hydrated lime [Ca(OH)₂]:** Produced by adding water to calcium oxide (CaO).

**Lime:** A term that is broadly applied in agriculture to identify materials that, when added to an acid soil, will neutralize soil acidity.

**Lime requirement:** That amount (expressed as amount per application area) of a liming material (normally agricultural limestone) required to change the soil water pH to a desired level based on soil characteristics and crop requirements.

**Marl:** Granular or loosely consolidated, earthy material composed largely of $CaCO_3$ in the form of seashell fragments; contains varying amounts of silt and organic matter.

**Neutralizing power:** Calculated on the basis of the chemical composition and fineness of a liming material.

**Pelletized lime:** A liming material (usually agricultural limestone) granulated into a pellet that eases handling and application, but does not alter the lime requirement recommendation or its effectiveness.

**Slag:** A by-product that has neutralizing capability, the more common being that from steel making, where limestone is used as a flux. The slag will contain various elements in a range concentrations depending on the process and elemental composition of the iron ore.

**Suspension lime:** Suspended finely ground aglime, 100-200 mesh, in 50-50 lime-water suspension.

## 18.2   LIMING MATERIALS

One could simply define a liming material as any substance that will correct soil acidity, and in some states, liming laws have such wording. What may be defined as a liming material or as aglime does not tell the whole story.

There has developed a "jargon" used to identify various liming materials that can be confusing. Limestone is a naturally occurring substance, that is, calcium carbonate $(CaCO_3)$. When identified as a liming material, $CaCO_3$ has been given various names: "agricultural limestone," or simply "agricultural lime," or "aglime," or just plain "lime." Another designation for only Ca-containing limestone is "calcitic limestone," or just plain "calcite," which separates only Ca-containing limestone from limestone that also contains Mg that would then be identified as either "dolomitic limestone," or just "dolomite." Therefore, one needs to be cautious when using these words as well as other identifying words for liming materials when specifying a particular liming material. The following are the common substances that are liming materials capable of correcting soil acidity:

1. *Limestone (CaCO₃):* Referred to as calcite limestone that exists in either an amorphous or crystalline form, depending on the conditions existing during its formation. These forms vary in their solubility, or reactivity, when brought into contact with an acid soil, the less crystalline being more reactive. Impurities in a limestone will also make the material more reactive, as pure limestone has a strong physical structure with the presence of impurities creating breaks in the atomic structure, thus rendering it more soluble.
2. *Dolomitic limestone:* Consists of a physical mixture of both calcium and magnesium carbonate, or as a double salt, calcium-magnesium carbonate, the more reactive or soluble being the double salt.
3. *Burned lime (CaO) and hydrate lime [Ca(OH)₂]:* Liming products that have high reactivity, and therefore must be used with caution when applied to an acid soil. They have use when a quick neutralizing effect is desired, but there will be little lasting effect after reaction with an acid soil. In addition, they are caustic and therefore require careful handling to prevent both physical and chemical injuries when being soil applied.
4. *Slags:* The by-products of various industrial and chemical processes, having chemical compositions reflecting their formation. Slags that are by-products of steel making are unique in that they contain sizable contents of elements that are essential to plants, such as Fe, Mn, Cu, and Zn, and in addition, may contain sizable amounts of both Mg and P. Used as a liming material, particularly on sandy soils, the accompanying elements may prove to be beneficial, making a slag by-product a more desirable liming source than other liming materials by contributing elements other than Ca and Mg.
5. *Pelletized Lime:* Limestone, sometimes finely ground (usually 100 mesh or smaller), is granulated into a pellet for ease of application, particularly for the home gardener or for the application of lime to small areas. Granulation does not change the liming effect, requiring the same lime rate as that for other forms of lime based on a determined lime requirement. Various substances are used to form the granular form, which does not impair the effectiveness of the lime being applied.

## 18.3 LIMING MATERIALS AND THEIR CALCIUM CARBONATE EQUIVALENTS (CCEs)

One of the quality factors for a liming material is its equivalency to that of pure calcium carbonate (CaCO₃), referred to as its calcium carbonate equivalent (CCE), with CaCO₃ having an equivalency of 100. All other liming materials are compared to that standard, with equivalencies greater than 100 being due to the compound's elemental composition and/or differences in the atomic weights of the primary elements in the compound. The following is a list of common aglime materials and their CCE percentages

| Aglime Material | CCE (%) |
|---|---|
| Calcium carbonate | 100 |
| Calcium limestone | 85–100 |
| Dolomitic limestone | 95–110 |
| Marl (Selma chalk) | 50–90 |
| Calcium hydroxide (slaked lime) | 120–133 |
| Calcium oxide (burnt or quick lime) | 150–175 |
| Calcium silicate | 86 |
| Basic slag | 50–70 |
| Ground oyster shells | 90–100 |
| Cement kiln dust | 40–100 |
| Wood ashes | 25–50 |
| Gypsum (land plaster) | None |
| By-products | Variable |

Liming materials, their calcium carbonate equivalents (CCEs), and sources are as follows:

| Name | Chemical Formula | Equivalent % $CaCO_3$ | Source |
|---|---|---|---|
| Calcium carbonate | $CaCO_3$ | 100 | Pure form, finely ground |
| Calcitic limestone | $CaCO3$ | 85–100 | Natural deposit |
| Dolomite | $CaCO_3+MgCO_3$ | 110 | Natural deposit |
| Shell meal | $CaCO_3$ | 95 | Natural shell deposits |
| Hydrated lime | $Ca(OH)_2$ | 120–135 | Steam burned |
| Burned lime | $CaO$ | 150–175 | Kiln burned |
| Calcium silicate | $CaSiO_3$ | 80–90 | Slag |
| Basic slag | $CaO$ | 50–70 | By-product of steel making |
| Power plant ash | $CaO, K_2O,$ $MgO, Na_2O$ | 5–50 | Wood-fired power plants |
| Cement kiln dust | $CaO, CaSiO_3$ | 40–60 | Cement plants |

The equivalent amounts of liming materials based on the CCE of the liming material compared to that for pure $CaCO_3$ are

| CCE of Liming Material (%) | Pounds Needed to Equal 1 Ton of Pure $CaCO_3$ |
|---|---|
| 60 | 3,333 |
| 70 | 2,857 |
| 80 | 2,500 |
| 90 | 2,223 |
| 100 | 2,000 |
| 105 | 1,905 |
| 110 | 1,818 |
| 120 | 1,667 |

The equivalent CCE for the liming material selected should be compared to that given for the liming material recommended. The danger here is that if there is a difference, that difference can result is either "under-liming" or "over-liming" the soil. It is not uncommon when liming an acid soil for the change in soil water pH to not either reach, or possibly exceed, the soil water pH desired.

## 18.4   MESH SIZE

A factor determining the liming effect is the mesh size particle distribution of the selected liming material. Liming materials are a mix of particle sizes, expressed as a percentage of the total that is able to pass screens in the 8- to 100-mesh range or even finer. The mesh size distribution required for marketing a liming material is set by statutes that vary by state based on what liming materials are commonly available. For example, Ag-ground limestone sold in Ohio must have the following mesh size distribution: 95% passing 8 mesh, 70% passing 20 mesh, 50% passing 60 mesh, and 40% passing 100 mesh; while in North Carolina, calcitic limestone particle size distribution required is 90% passing 20 mesh and 25% passing 100 mesh.

Mesh size distribution will also vary, depending on whether calcitic or dolomitic limestone, as dolomite is less water soluble than calcite, dolomitic limestone will have a greater percentage of finer material than that for calcitic limestone. In North Carolina, calcitic limestone must have a particle distribution of 90% passing 20 mesh and 25% passing 100 mesh, while dolomitic limestone requires a particle size distribution of 90% passing 20 mesh and 35% passing 100 mesh.

Another category, pulverized or super fine, comprises liming materials that have a high percentage of fine particles, pulverized limestone with 100% passing 8 mesh, 95% passing 20 mesh, 70% passing 60 mesh, and 60% passing 100 mesh; while the super-fine will have 100% passing both 8 and 20 mesh, 95% passing 60 mesh, and 60% passing 100 mesh.

Liming materials with a particular mesh size will react differently in an acid soil, which can be of significant value under various cropping conditions. In the following table, the percentage of applied lime that has reacted with time based on mesh size is shown

|  | Years after Application | |
| --- | --- | --- |
| Mesh Size | 1 | 2 |
|  | % Reacted | |
| Coarser than 8 | 5 | 15 |
| 8–20 | 20 | 45 |
| 20–50 | 50 | 100 |
| 50–100 | 100 | 100 |

For some situations, selecting a liming material with either a majority of coarse or fine particles can be used to establish a time factor in the liming effect.

## 18.5 QUALITY FACTOR DESIGNATION

The quality of a liming material is based on its chemical composition, which determines the Calcium Carbonate Equivalent (CCE), and fineness of grind. State laws govern what can be designated and sold as a liming material (usually defined as a substance that, when applied to an acid soil, will neutralize soil acidity), with minimum percent distribution ranges for particle sizes by mesh group. Liming materials are periodically tested to ensure that marketed liming materials conform to state laws and the given label properly identifies the substance.

## 18.6 LIME REQUIREMENT (LR)

A lime requirement (LR) determination specifies the quantity of a liming material needed to increase the pH of an acid soil to a particular pH level. The LR can be estimated based on the textural class, organic matter content or soil type, or a combination of these soil properties. The quantity of lime needed to adjust the soil pH based on soil texture is given in the following table:

**Approximate Amount of Finely Ground Limestone Needed to Raise the pH of a 7-inch Layer of Soil**

| Soil Texture | Lime Requirement (lbs/1,000 sq. ft.) | |
|---|---|---|
| | From pH 4.5 to 5.8 | From pH 5.5 to 6.5 |
| Sand and loamy sand | 23 | 28 |
| Sandy loam | 37 | 60 |
| Loam | 55 | 78 |
| Silt loam | 69 | 92 |
| Clay loam | 87 | 106 |
| Clay | 174 | 107 |

*Source:* From *USDA Agricultural Handbook No. 18.*

Note that the liming material, desired soil pH to be obtained, and the depth of application are specified.

From another source, a similar estimation of lime requirement based on the soil water pH and textural class is shown in this table:

**Quantity of Ag-Ground Limestone (TNP = 90, fineness of 40% <100-mesh, 50% <60-mesh, 95% <8-mesh) Required (1,000 lbs/A) to Raise an Acid Soil pH to 6.5 to a depth of 6-2/3 inches based on Soil Water pH and Soil Texture (S-silt; SL- silty loam; L – loam; SiCl – silty clay; O – organic soil)**

| Soil Water pH | Lime Requirement (1000 lbs/A) | | | | |
|---|---|---|---|---|---|
| | S | SL | L | SiCl | O |
| 3.5–3.9 | 4.0 | 6.5 | 9.0 | 12.0 | 20.0 |
| 4.0–4.4 | 3.0 | 5.0 | 7.0 | 9.0 | 15.0 |
| 4.5–4.9 | 2.0 | 3.5 | 5.5 | 6.5 | 10.0 |

| | | | | | |
|---|---|---|---|---|---|
| 5.0–5.4 | 1.5 | 2.5 | 4.0 | 5.0 | 5.0 |
| 5.5–5.9 | 1.0 | 1.5 | 2.5 | 3.5 | nr |
| 6.0–6.4 | 0.5 | 1.0 | 1.5 | 2.5 | nr |

*Note:* nr = none required.

The quantity of lime to be applied is specified based on the type of lime (TNP and fineness) and mixing depth. Any change of either would require a change in the amount applied.

Today, the LR is determined using a combination of soil water and buffer pH values. The buffer pH can be determined by (1) an acid-base titration, or (2) by the use of one of several buffer procedures selected on the basis of the physiochemical properties of the soil (see Section II, "Physiochemical Properties of Soil"). Lime recommendation tables are specifically based on soil type and desired pH adjustment, as well as crop to be grown and depth of lime incorporation. Adjustments are also made based on lime quality factors, such as CCE (TNP) and mesh size.

## 18.7 SOIL TEST RATIO OF Ca TO Mg DETERMINES FORM OF LIMESTONE TO APPLY

The form of limestone to be applied, either calcitic or dolomitic, can be based on the soil test level of both Ca and Mg. For soils "low" (see page ) in soil test Mg, it favors the use of dolomitic limestone, while either form could be selected if neither Ca nor Mg were soil testing "low." Combining the ratio of soil test Ca to Mg with the lime requirement, that form of limestone would be designated as shown in the following table:

| Ca:Mg Soil Ratio | Lime Requirement (lbs/A) | Form of Limestone |
|---|---|---|
| >10:1 | >4,000 | All dolomitic |
| <10:1 | >4,000 | All calcitic |
| >10:1 | <4,000 | 60:40 dolomitic:calcitic |
| <10:1 | <4,000 | 40:60 dolomitic:calcitic |

## 18.8 LIMING RATE DETERMINED BY ACIDIFYING EFFECT OF FERTILIZER

A lime recommendation can be adjusted to compensate for the acidifying effect from applied N fertilizers as given in the following table:

| Nitrogen Fertilizer | Amount of CaO to Compensate the Soil Acidification Induced by 2.2 lbs N |
|---|---|
| Calcium ammonium nitrate | 1.32 lb |
| Ammonia, urea, ammonium nitrate | 2.2 lb |
| Diammonium phosphate | 4.4 lb |
| Ammonium sulfate | 6.6 lb |

*Source: IFA World Fertilizer Use Manual, 1992.*

## 18.9   LIME SHOCK

When liming an acid soil, depending on the properties of the liming material and the amount applied, plus the physiochemical properties of the soil, there can occur a temporary adverse reaction that can affect an established crop, the term "lime shock" being applied to such an occurrence. Therefore, it is generally recommended that a liming material be applied at least 4 months prior to planting a crop, allowing sufficient time for the lime-soil interaction to stabilize. If the lime requirement specifies that a sizable quantity of liming material is required, dividing the requirement into equal portions and allowing sufficient time for reaction before the next application may be desirable to minimize the "lime shock" effect.

## 18.10   LIME INCORPORATION

The incorporation of a liming material into a soil can be a challenge when applied to an already growing crop, when minimum tillage practices are in use, and when land preparation procedures do not result in a mixing within the normal rooting depth. When a long-term crop will be established, the selection of a liming material (solubility and mesh size distribution) should ensure a long-term liming effect. Between cropping sequences, deep plowing and incorporation of a selected liming material can ensure some degree of constancy over the cropping period. Because most liming materials are not "soluble," their incorporation into the soil, if surface applied, will be very slow, creating a layering of the soil water pH with depth that is not conducive to good plant growth.

## 18.11   DEPTH OF INCORPORATION

Normally, a lime requirement recommendation specifies the depth of incorporation that may be designated as just to the "plow depth," or to an actual specified depth. If the depth of incorporation will not be that specified for the recommended rate, then the amount of lime applied should be adjusted to avoid either an under- or over-application. To make a depth adjustment from a recommended 7-inch depth of incorporation, the following multiplying factors are used:

### Depth of Incorporation

| Inches | Multiplying Factor |
| --- | --- |
| 3 | 0.43 |
| 4 | 0.57 |
| 5 | 0.71 |
| 6 | 0.86 |
| 7 | 1.00 |
| 8 | 1.14 |
| 9 | 1.29 |
| 10 | 1.43 |
| 11 | 1.57 |

## 18.12   SUBSOIL pH

What is frequently forgotten is the pH of the subsoil, that portion of the soil profile below the normal rooting depth of most plants. The pH of the subsoil may be equally important as that of the surface soil for some crop plants and cropping systems if roots venture into the subsoil, adding to the supply of water and essential plant elements available to the plant. The extent of penetration and development will depend on both the physical and physiochemical properties of the subsoil. Normally with time, maintaining the surface soil (plow depth or the normal plant rooting depth) at the optimum pH will tend to keep the pH of the subsoil constant. Allowing the surface soil pH to cycle from acidity to near neutral by intermittent liming will tend to make the subsoil more acidic with time. With the ability to deep plow and apply a liming material within or at the surface of the subsoil can be used to keep the subsoil pH at that level conducive for good root development. If crop performance is less than expected when best management practices are employed, soil testing the subsoil may be helpful in identifying subsoil acidity as a possible cause of poor crop performance.

# 19 Inorganic Chemical Fertilizers and Their Properties

Inorganic chemical fertilizers are essential for successful crop production, correcting soil fertility insufficiencies, and providing essential plant nutrient elements.

## 19.1 DEFINITIONS

**Fertilizer:** A substance containing one or more of the essential plant elements that, when added to a soil/plant system, aids plant growth and/or increases productivity by providing additional essential elements for plant use.

**Fertilizer elements:** The three essential major plant elements N, P, and K are so identified because they are the commonly included elements in a wide range of fertilizer materials.

**Fertilizer grade:** A term used to identify a mixed fertilizer containing two or three of the fertilizer elements in the order, N, P expressed as its oxide ($P_2O_5$), and K expressed as its oxide ($K_2O$), the content of each element given in percent.

## 19.2 FERTILIZER TERMINOLOGY

**Acid-forming fertilizer:** Fertilizer, after application to and reaction with the soil, increases the residual acidity and decreases the soil pH.

**Analysis:** Determination by chemical means and expressed in terms that laws permit of the elemental content of a fertilizer material. Although *analysis* and *grade* sometimes are used synonymously, *grade* applies only to the three primary plant essential elements (N, available $P_2O_5$, and soluble $K_2O$) and is stated as the guaranteed minimum quantity present.

**Banded fertilizer:** Placement of fertilizer in a concentrated zone either on or below the soil surface.

**Blended:** A mechanical mixture of different fertilizer materials.

**Broadcast:** Application either of a solid or fluid fertilizer on the soil surface with or without subsequent incorporation by tillage.

**Brand:** Term, designation, or trademark used in connection with one or several grades of fertilizer; trade name assigned by the manufacturer to its fertilizer product.

**Bulk:** Delivered to the user in an unpackaged solid or liquid form in a large lot.

**Bulk-blended:** A physical mixture of dry granular fertilizer materials to produce specific fertilizer ratios and grades.

**Complete:** A chemical compound or a blend of compounds containing significant quantities of N, P, and K. It may also contain other plant essential elements.

**Compound:** A fertilizer formulated with two ore more plant essential elements.

**Controlled release:** A fertilizer term used interchangeably with delayed release, slow release, controlled availability, slow acting, and metered release to designate a controlled dissolution of fertilizer at a lower rate than conventional water-soluble fertilizer. Controlled release properties either result from coatings on water-soluble fertilizer or from low dissolution and/or mineralization rates of fertilizer materials in soil.

**Dribble fertilization:** Dribbling or strip banding fertilizer in a form of band placement that involves application of solid or fluid fertilizers in bands or strips of varying widths on the soil surface or on the surface of crop residues.

**Dual placement:** Simultaneous placement of two fertilizer materials in subsurface bands.

**Elements:** Refers to the major elements, N, P and K, being the main elemental ingredients in many commonly used fertilizers.

**Fertigation:** Application of fertilizer in irrigation water.

**Fertilizer use efficiency:** An expression of the units of yield per unit of nutrient provided for the crop.

**Filler:** A substance added to fertilizer materials to provide bulk, prevent caking, or serve some purpose other than providing essential plant essential elements.

**Fluid:** Fertilizer wholly or partially in solution that can be handled as a liquid, including clear liquids and liquids containing solids in suspension.

**Foliar:** Application of a dilute solution of an essential plant nutrient element (confined primarily to the micronutrients) to plant foliage usually made to supplement essential plant nutrient elements absorbed by plant roots.

**Formula:** Designation of quantities of ingredients combined to make a fertilizer.

**Grade:** The guaranteed minimum analysis in percent of the major plant essential elements (N, P, and K expressed as $N–P_2O_5–K_2O$ in that series) contained in a fertilizer material or in a mixed fertilizer.

**Granular:** Fertilizer in the form of particles sized between an upper and lower limit or between two screen sizes, usually within the range of 1 to 4 mm and often more closely sized.

**Hydroscopic:** Materials that absorb moisture from the atmosphere and, as a result, will cake and not evenly flow. Substances may be added to prevent moisture absorption.

**Injected:** Placement of fluid fertilizer or anhydrous ammonia ($NH_3$) into the soil either through the use of pressure or non-pressure systems.

**Inorganic:** Fertilizer materials that do not contain carbon (C). Urea [$CO(NH_2)_2$], although containing C, is normally considered an inorganic fertilizer material.

**Liquid:** A fluid in which the ingredients are in true solution.

**Mineral:** A term interchangeably used with *inorganic*, although more specifically refers to those substances or an ingredient in a fertilizer that is a natural occurring substance, such as potassium chloride (KCl).

**Mixed:** Two or more fertilizer materials blended or granulated together into individual mixes.

**Organic:** A material or materials that contain carbon (C), in addition to hydrogen (H) and oxygen (O), generally applying to those products derived from plant and animal sources, such as manures, plant by-products, or from natural sources (peat, bark, etc.).

**Point injection:** Use of a spoked wheel to inject fluid fertilizer into the rooting zone.

**Pop-up:** Fertilizer placed in small amounts in direct contact with the seed (a method of limited use and frequently of no benefit).

**Preplant fertilizer:** Fertilizer applied to the soil prior to planting.

**Primary elements:** Refers to the three essential elements, N, P, and K, required by plants in fairly large quantities, and are the major elements in mixed and grade-designated fertilizers, expressed as N, $P_2O_5$, and $K_2O$.

**Ratio:** The numerical relationship of concentrations of the primary elements (NPK), for example a 5-10-15 fertilizer grade would have 1:2:3 ratio.

**Requirement:** Quantity of certain plant nutrient elements needed, in addition to the amount supplied by the soil, to increase plant growth to a designated level.

**Salt-index:** Used to compare solubilities of chemical compounds used as fertilizers. Most N and K compounds have high indices, and P compounds have low indices. High index compounds applied in direct seed contact at too high rates can cause seedling damage because the compounds have a high infinity for water.

**Side-banded fertilizer:** Application of fertilizer in bands on one side or both sides of the seedlings.

**Split application:** Fertilizer applied two or more times during the crop-growing season. Preplant and one or more post-plant applications are common.

**Strip fertilizer:** Fertilizer applied in surface bands that may be incorporated by tillage or remain on the soil/residue surface.

**Starter:** Fertilizer applied in relatively small amounts with or near the seed for the purpose of accelerating early growth of the crop plant.

**Suspension or slurry:** Fluid fertilizer containing dissolved and undissolved plant essential elements, kept in suspension with a suspending agent, usually a swelling-type clay. Suspensions are flowable enough to be mixed, pumped, agitated, and applied to the soil as a homogeneous mixture.

**Top-dressed:** Surface application of fertilizer to a soil after the crop has been established.

**Unit:** A unit of a plant essential element is percent of a ton. A ton of a 6-12-4 fertilizer would contain 6 units of N, 12 units of $P_2O_5$, and 4 units of $K_2O$.

## 19.3   CHARACTERISTICS OF THE MAJOR ELEMENTS AS FERTILIZER

1. *Nitrogen (N):* The form of N applied in inorganic form will be either as the ammonium cation ($NH_4^+$) or the nitrate anion ($NO_3^-$), or the combination of both ($NH_4NO_3$), ions that have completely different behaviors in soil. Urea [$CO(NH_2)_2$], which contains neither ammonium nor nitrate, will be quickly hydrolyzed when brought into contact with a soil, the hydrolyzed N form being ammonium. Urea left on the soil surface will hydrolyze with the production of gaseous ammonia ($NH_3$) that will be released into the atmosphere, resulting in a significant loss of applied N. The $NH_4^+$ cation will participate in the cation exchange phenomenon of the soil, and that in the soil solution will undergo nitrification into the $NO_3^-$ anion. The $NH_4^+$ cation moves in the soil solution primarily by diffusion. The $NO_3^-$ anion moves readily within the soil by mass flow and can be leached from the rooting profile by rainfall or applied irrigation water.

2. *Phosphorus (P):* The form of P applied as fertilizer (see page 159) may be as the orthophosphate anion ($PO_4^{3-}$), the diphosphate anion ($H_2PO_4^-$), or the monophosphate anion ($HPO_4^{2-}$), depending on the fertilizer chemical composition. Phosphorus availability for root absorption is a complex process of interacting soil chemistries. Even though a highly soluble form of P is applied, such as phosphoric acid ($H_3PO_4$), the orthophosphate anion ($PO_4^{3-}$) will quickly interact with the soil, forming precipitates with the elements Ca, Fe, and Al, and that remaining in soluble form will exist in the soil solution as either the diphosphate anion ($H_2PO_4^-$) or as the monophosphate anion ($HPO_4^{2-}$), depending on the pH of the soil, and moves primarily in the soil solution by diffusion.

3. *Potassium (K):* The K form in fertilizer is as the potassium cation ($K^+$), and when in solution participates in the cation exchange phenomenon of the soil. The $K^+$ cation moves in the soil primarily by mass flow.

4. *Calcium (Ca):* The Ca form applied as a liming material (see page 148) or as a fertilizer (see Table 19.1) is as the calcium cation ($Ca^{2+}$), and when in solution participates in the cation exchange phenomenon of the soil. The $Ca^{2+}$ cation moves in the soil primarily by mass flow.

5. *Magnesium (Mg):* The Mg form applied as a liming material (see page 148) or as a fertilizer (see Table 19.2) is as the magnesium cation ($Mg^{2+}$), and when in solution participates in the cation exchange phenomenon of the soil. The $Mg^{2+}$ cation moves in the soil solution primarily by diffusion.

6. *Sulfur (S):* The S form in fertilizers is as the sulfate anion ($SO_4^{2-}$), and when in solution interacts with other elements, mainly Ca and Al (existing as cations in the soil solution), to form precipitates. The $SO_4^{2-}$ anion moves in the soil solution primarily by diffusion.

## 19.4   CONVERSION FACTORS FOR THE MAJOR ESSENTIAL FERTILIZER ELEMENTS

- To convert $NH_3$ to elemental N, multiply by 0.8226.

## TABLE 19.1
## Primary Fertilizer Sources for the Major Elements, Formula, Form, and Percent Content

### Nitrogen (N)

| Source | Formula | Form | %N |
|---|---|---|---|
| *Inorganic* | | | |
| Ammonium nitrate | $NH_4NO_3$ | Solid | 34 |
| Ammonium sulfate | $(NH_4)_2SO_4$ | Solid | 21 |
| Anhydrous ammonia | $NH_3$ | Gas | 82 |
| Aqua ammonia | $NH_4OH$ | Liquid | 2–25 |
| Nitrogen solutions | Varies | Liquid | 19–32 |
| Monoammonium phosphate | $NH_4H_2PO_4$ | Solid | 11 |
| Diammonium phosphate | $(NH_4)_2HPO_4$ | Solid | 1–18 |
| Calcium nitrate | $Ca(NO_3)_2 \cdot 4H_2O$ | Solid | 16 |
| Potassium nitrate | $KNO_3$ | Solid | 13 |
| *Synthetic organic* | | | |
| Urea | $CO(NH_2)_3$ | Solid | 45–46 |
| Sulfur-coated urea | $CO(NH_2)_2$-S | Solid | 40 |
| Urea-formaldehyde | $CO(NH_2)_2$-$CH_2O$ | Solid | 38 |
| *Natural Organic* | | | |
| Cotton seed meal | | Solid | 12–13 |
| Animal manure | | Solid | 10–12 |
| Sewage sludge | | Solid | 10–20 |
| Chicken litter | | Solid | 20–40 |

### Phosphorus (P)

| Source | Formula | Form | % Available $P_2O_5$ | |
|---|---|---|---|---|
| | | | Citrate soluble | Water soluble |
| Superphosphate, 0-20-0 | $Ca(H_2PO_4)_2$ | | Solid 16–20 | 90 |
| Conc superphosphate, 0-45-0 | $Ca(H_2PO_4)_2$ | | Solid 44–52 | 92–98 |
| Monoammonium phosphate | $NH_4H_2PO_4$ | | Solid 48 | 100 |
| Diammonium phosphate | $(NH_4)_2HPO_4$ | | Solid 46–48 | 100 |
| Ammonium polyphosphate | $(NH_4)_2HP_2O_7 \times H_2O$ | | Solid 34 | 100 |
| Phosphoric acid | $H_3PO_4$ | | Liquid 55 | 100 |
| Rock phosphate, fluor- and chloroapatite | $3Ca_4(PO_4)_2CaF_2$ | | Solid 3–26 | — |
| Basic slag | $5CaO-P_2O_5SiO_2$ | | Solid 2–16 | — |
| Bone meal | — | | Solid 22–28 | — |

### Potassium (K)

| Source | Formula | Form | %K |
|---|---|---|---|
| Potassium chloride, muriate of potash | KCl | Solid | 60–61 |

(*continued*)

**TABLE 19.1 (*continued*)**
**Primary Fertilizer Sources for the Major Elements, Formula, Form, and Percent Content**

### Potassium (K)

| Source | Formula | Form | %K |
|---|---|---|---|
| Potassium sulfate | $K_2SO_4$ | Solid | 50–52 |
| Potassium magnesium sulfate (SUL-PO-MAG) | $K_2SO_4 \times MgSO_4$ | Solid | 22 |
| Potassium nitrate | $KNO_3$ | Solid | 44 |

### Calcium (Ca)

| Source | Formula | Form | %Ca |
|---|---|---|---|
| Calcium nitrate | $Ca(NO_3)_2 \cdot 4H_2O$ | Solid | 19 |
| Superphosphate, 0-20-0 normal | $Ca(H_2PO_4)_2 + CaSO_4 \times 2H_2O$ | Solid | 20 |
| Superphosphate, 0-45-0 triple | $Ca(H_2PO_4)_2$ | Solid | 14 |
| Gypsum | $CaSO_4 \cdot 2H_2O$ | Solid | 23 |
| Gypsum, by-product | $CaSO_4 \cdot 2H_2O$ | Solid | 17 |

### Magnesium (Mg)

| Source | Formula | Form | %Mg |
|---|---|---|---|
| Kieserite, magnesium sulfate | $MgSO_4 \cdot 7H_2O$ | Solid | 18 |
| Epsom salts, magnesium sulfate | $MgSO_4 \cdot 7H_2O$ | Solid | 10 |
| Potassium magnesium sulfate | $K_2SO_4 \cdot MgSO_4$ | Solid | 11 |
| Magnesium oxide | $MgO$ | Solid | 50–55 |

### Sulfur (S)

| Source | Formula | Form | %S |
|---|---|---|---|
| Sulfur (elemental) | S | Solid | 90-100 |
| Ammonium sulfate | $(NH_4)_2SO_4$ | Solid | 24 |
| Gypsum (land plaster) | $CaSO_4 \cdot 2H_2O$ | Solid | 19 |
| Magnesium sulfate, Epsom salts | $MgSO_4 \cdot 7H_2O$ | Solid | 13 |
| Potassium sulfate | $K_2SO_4$ | Solid | 18 |
| Potassium magnesium sulfate (SUL-PO-MAG) | $K_2SO_4 \cdot MgSO_4$ | Solid | 23 |
| Superphosphate, 0-20-0 | $CaSO_4$ + Calcium phosphate | Solid | 12 |
| Ammonium thiosulfate | $(NH_4)_2S_2O_3$ | Solid | 26 |
| Sulfur-coated urea | $CO(NH_2)_2$-S | Solid | 10 |

## TABLE 19.2
## Primary Fertilizer Sources for the Micronutrients, Formula, Form, and Percent Content

### Boron (B)

| Name | Formula | % B |
|---|---|---|
| Fertilizer Borate, 48 | $Na_2B_4O_7 \cdot 10H_2O$ | 14–15 |
| Fertilizer Borate, granular | $Na_2B_4O_7 \cdot 10H_2O$ | 14 |
| Foliarel | $Na_2B_8O_{13} \cdot 4H_2O$ | 21 |
| Solubor | $Na_2B4O_7 \cdot 4H_2O + Na_2B_{10}O_{16} \times 10H_2O$ | 20 |
| Borax | $Na_2B_4O_7 \cdot 10H_2O$ | 11 |

### Chlorine (Cl)

| Name | Formula | % Cl |
|---|---|---|
| Potassium chloride | $KCl$ | 47 |

### Copper (Cu)

| Name | Formula | %Cu |
|---|---|---|
| Copper sulfate (monohydrate) | $CuSO_4 \cdot H_2O$ | 35 |
| Copper sulfate (pentahydrate) | $CuSO_4 \cdot 5H_2O$ | 25 |
| Cupric oxide | $CuO$ | 75 |
| Cuprous oxide | $Cu_2O$ | 89 |
| Cupric ammonium phosphate | $Cu(N_4)PO_4 \cdot H_2O$ | 32 |
| Basic copper sulfates | $CuSO_4 \cdot 3Cu(OH)_2$ (general formula) | 13–53 |
| Cupric chloride | $CuCl_2$ | 17 |
| Copper chelates | $Na_2CuEDTA$ | |
| | $NaCuHEDTA$ | |
| | Organically bound Cu | 5–7 |

### Iron (Fe)

| Name | Formula | %Fe |
|---|---|---|
| Ferrous ammonium phosphate | $Fe(NH_4)PO_4 \cdot H_2O$ | 29 |
| Ferrous ammonium sulfate | $(NH_4)_2SO_4 \cdot FeSO_4 \cdot 6H_2O$ | 14 |
| Ferrous sulfate | $FeSO_4 \cdot 7H_2O$ | 19–21 |
| Ferric sulfate | $Fe(SO_4)_3 \times 4H_2O$ | 23 |
| Iron chelates | $NaFeEDTA$ | 5–11 |
| | $NaFeHFDTA$ | 5–9 |
| | $NaFeEDDHA$ | 6 |
| | $NaFeDTPA$ | 10 |
| Iron polyflavonoids | Organically bound Fe | 9–10 |

### Manganese (Mn)

| Name | Formula | %Mn |
|---|---|---|
| Manganese sulfate | $MnSO_4 \cdot 4H_2O$ | 26–28 |

*(continued)*

**TABLE 19.2** (*continued*)
**Primary Fertilizer Sources for the Micronutrients, Formula, Form, and Percent Content**

### Manganese (Mn)

| Name | Formula | %Mn |
|---|---|---|
| Manganese oxide | MnO | 41–68 |
| Manganese chelate | MnEDTA | 5–12 |

### Molybdenum (Mo)

| Name | Formula | %Mo |
|---|---|---|
| Ammonium molybdate | $(NH_4)_6Mo_7O_{24} \cdot 2H_2O$ | 54 |
| Sodium molybdate | $Na_2MoO_4 \cdot 2H_2O$ | 39–41 |
| Molybdenum trioxide | $MnO_2$ | 66 |

### Zinc (Zn)

| Name | Formula | %Zn |
|---|---|---|
| Zinc sulfate | $ZnSO_4 \cdot 7H_2O$ | 35 |
| Zinc oxide; | ZnO | 78-80 |
| Zinc chelates | $Na_2ZnEDTA$ | 14 |
|  | NaZnTA | 13 |
|  | NaZnHEDTA | 9 |
| Zinc polyflavonoids | Organically bound Zn | 10 |

- To convert $K_2O$ to elemental K, multiply by 0.8301
- To convert elemental K to $K_2O$, multiply by 1.2046
- To convert $P_2O_5$ to elemental P, multiply by 1.2046
- To convert elemental P to $P_2O_5$, multiply by 2.2912

## 19.5 CHARACTERISTICS OF THE MICRONUTRIENTS AS FERTILIZER

1. *Boron (B):* Exists primarily in the soil organic matter, and because it exists in soluble forms will easily move with soil water, being leached from the rooting profile by heavy rain and excessive irrigation, and will move to the surface of a soil with water evaporation from the soil surface.
2. *Chlorine (Cl):* Exists in the soil as the chloride ion ($Cl^-$) that easily moves with soil water. Being an element that is ever-present in the environment, a contaminate in most fertilizers, and is added with the use of potassium chloride (KCl) as a major fertilizer source for K.
3. *Copper (Cu):* Exists in the soil in both organic and inorganic forms, and being a cation ($Cu^{2+}$), participates in the cation exchange phenomenon. The $Cu^{2+}$ cation moves in the soil solution primarily by diffusion.
4. *Iron (Fe):* Exists in the soil in both organic and inorganic forms, and being a cation [ferrous ($Fe^{2+}$) or ferric ($Fe^{3+}$)], depending on the oxidation state of

the soil environment, participates in the cation exchange phenomenon. The $Fe^{2+}$ and $Fe^{3+}$ cations move in the soil solution primarily by diffusion.

5. *Manganese (Mn):* Exists in the soil in both organic and inorganic forms, and being a cation ($Mn^{2+}$), participates in the cation exchange phenomenon. The $Mn^{2+}$ cation moves in the soil solution primarily by diffusion.

6. *Molybdenum (Mo):* Exists in the soil as an anion ($MoO_4^-$) and moves within the soil solution primarily by diffusion.

7. *Zinc (Zn):* Exists in the soil in both organic and inorganic forms, and being a cation ($Zn^{2+}$), participates in the cation exchange phenomenon. The $Zn^{2+}$ cation moves in the soil solution primarily by diffusion.

## 19.6 THE PHYSICAL AND CHEMICAL PROPERTIES OF FERTILIZERS

### 19.6.1 INORGANIC

The reactivity of a fertilizer, when brought into contact with soil, is determined by the physical and chemical properties of both the fertilizer and the soil, the fertilizer factors being

- Physical form (solid, liquid, gas)
- Solubility
- Elemental composition
- Chemical reactivity
- Method of application
- Time of application

and soil factors being

- pH
- Texture
- Organic matter content
- Existing elemental levels

### 19.6.2 FERTILIZER FACTORS

The physical form of the fertilizer will determine the method of soil application, how it is stored, and also what equipment is required for its application.

Solubility can be both a positive and/or negative factor in its reaction with the soil and the availability of the applied essential plant element for root absorption. High solubility may actually decrease element availability due to its fast reaction with the soil, although enhancing movement by diffusion (see page 12) within the soil solution.

Elemental composition can mean that more than one essential plant nutrient element can be applied by the use of a single fertilizer material. This may be a beneficial factor, but it also may limit the use of a particular fertilizer when one of the accompanying elements is not needed, and with repeated use, result in an imbalance of that element in the soil, or that additional element may be needed in greater quantity than that being applied based on the primary element recommendation.

Chemical reactivity may alter the elemental balance in the soil as the result of precipitation and by altering the cation exchangeable makeup on the soil colloids, and/or alter the soil pH from generated acidity (see page 57).

Method of soil application can significantly influence the availability for root absorption of elements in the fertilizer, and/or affect the distribution of an element within the rooting zone. Broadcasting and mixing by tilling into the soil will disperse the fertilizer particles, affecting availability and fixation processes, while banding the fertilizer will reduce soil–fertilizer particle contact, reduce the potential for soil fixation, and increase elemental availability, thereby increasing the absorption potential by contacting roots. Repeated banding of fertilizer will eventually create zones of high and low elemental availability that will impact crop performance. Row banding at planting and row side dressing during plant growth are common practices for maximizing root absorption and minimizing soil fixation. Fertigation, applying fertilizer through irrigation water (see page 173), has particular use for specialty crops when precise timing and rate of application of an essential plant nutrient element will benefit plant growth at critical periods, thereby increasing both fruit yield and quality.

Time of fertilizer application will significantly affect plant response—the lesser the time between application and plant need, the more positive the effect; the longer the period between application and planting, the lesser the effect.

### 19.6.3   SOIL FACTORS

Both acidity and alkalinity will influence the availability of applied essential plant elements, keeping in mind that the chemical and physical properties of the fertilizer will also be factors, thus making the interaction of equal importance.

Soil texture (see page 49) determines the extent of the reactive soil particle surfaces; the coarser (sandy) the soil, the lower the size of reactive surfaces, the heavier (clay) the soil, the greater the size of the reactive surface. The nature of the colloidal material (see page 54) will also determine the level of reactivity between these soil particles and fertilizer particles.

As the soil organic matter content increases, there is greater potential for reactivity between the applied fertilizer and soil. The nature of the organic factor in the soil is also a significant factor (see page 65).

The existing elemental equilibrium within the soil will be altered with the addition of that element as fertilizer. Equilibrium systems vary with the element; therefore, the shift in equilibrium will be element determined. For example, applying a K-containing fertilizer will shift the existing equilibrium that exists between K in the soil solution, that on the cation exchange complex, and that in various fixed forms. The rapidity of this equilibrium shift will be determined by the physiochemical properties of the soil, the soil moisture level, and the presence or absence of an active growing crop.

## 19.7  NATURALLY OCCURRING INORGANIC FERTILIZERS

### 19.7.1  ROCK PHOSPHATE

Rock phosphate, a commonly used P-source fertilizer in the past, is now coming back into use as an acceptable P source when growing organically; its reactivity in soil will be determined by:

- Particle size (fineness)
- Method of application, broadcast versus banding
- Soil pH, greater P availability with increased acidity
- Soil organic content, decreased P availability with higher OM content
- Soil texture, higher P availability in coarse (sandy) textured soils

The finer the particle size, the greater will be rock phosphate's reactivity, which can work both ways, either increasing (more water soluble with large surface area) or even decreasing P availability to a growing crop due to chemical reaction with the soil. Soil pH, organic matter content, and soil texture will affect rock phosphate P availability. Method of application will determine its solubility and P availability. Broadcast application provides high contact with the soil, which in turn, will reduce P availability. Applying rock phosphate in bands will increase P availability when roots make contact with the bands. Rock phosphate is a desired P source due to its long-term presence in the soil, slowly releasing P with the weathering process that occurs in a cropped soil.

### 19.7.2  POTASSIUM CHLORIDE (KCl) AND POTASSIUM SULFATE ($K_2SO_4$)

Both of these substances are naturally occurring, KCl mined as a natural occurring mineral, while $K_2SO_4$ is found in various salt brines found in semi-arid and arid regions of the world.

### 19.7.3  LIMESTONE

Limestone, calcitic and dolomitic limestone, are natural substances, calcite providing an essential plant nutrient element (Ca) and dolomite providing the essential plant nutrient elements Ca and Mg.

# 20 Organic Fertilizers and Their Properties

Organic farming/gardening (see Chapter 25) requires the use of organic or naturally occurring substances to supply the essential plant nutrient elements. Some of these materials are readily available, while others have physical and/or chemical properties that limit their acceptability and use. Organic fertilizers vary considerably in elemental content and elemental form, whether in either inorganic soluble or organic insoluble form.

## 20.1 VALUE

One of the desired values of some organic fertilizers is that they also contribute to the organic components of the soil, thereby improving friability, water-holding capacity, and upon decomposition, contributing to the cation exchange capacity of the soil with the formation of humus, a by-product of organic matter decomposition, as well as adding plant nutrient elements that will contribute to the soil fertility status.

## 20.2 COMPOSTED ANIMAL MANURES

Composted manures and similar organic substances are better suited for use than what would be termed fresh, in terms of their physical and chemical stabilities, and freedom from disease organisms and weed seeds. With confined animals, manures frequently contain litter and/or bedding materials, organic and inorganic, that are not evenly distributed when the manure is gathered for disposal. Composting the gathered manure containing litter/bedding materials acts to unify its composition and stabilize its chemical and physical characteristics. For use as a soil amendment, composted manures may be either air or oven dried, and then ground to form a uniform and easily handled material.

## 20.3 ANIMAL MANURE MAJOR ELEMENT COMPOSITION

Elemental content values for most animal manures are well known, with typical composition values for the fertilizer elements, N, P, and K, with P and K expressed as their oxides, $P_2O_5$ and $K_2O$, as follows:

## Typical Compositions of Manures

| Source | Dry Matter | Approximate Composition[a] | | |
|---|---|---|---|---|
| | | N | P$_2$O$_5$ | K$_2$O |
| | (%) | (% dry weight) | | |
| Dairy | 15–26 | 0.6–2.1 | 0.7–1.1 | 2.4–3.6 |
| Feedlot | 20–40 | 1.0–2.5 | 0.9–1.6 | 2.4–3.6 |
| Horse | 15–25 | 1.7–3.0 | 0.7–1.2 | 1.2–2.2 |
| Poultry | 20–30 | 2.0–4.5 | 4.5–6.0 | 1.2–2.4 |
| Sheep | 25–35 | 3.0–4.0 | 1.2–1.6 | 3.0–4.0 |
| Swine | 20–30 | 3.0–4.0 | 0.4–0.6 | 0.5–1.0 |

[a] The range in N-P$^2$O$^5$-K$^2$O element content shown above reflects the influence that the type of feed consumed, the percentage and type of litter or bedding materials used, and the age and degree of decomposition has on final composition. Also note that the dry matter content of these manures varies considerably, which, in turn, will affect its elemental composition. The extent of aeration during composting will determine the extent of decomposition of extraneous material and the degree of N preservation.

Land disposal of animal manures can pose significant environmental concerns if after application there is substantial runoff, with N dissolved or suspended in the runoff water and with P being primarily suspended with the soil materials being carried into streams and lakes. Even groundwater contamination can occur when animal manures and composts are frequently applied to crop production lands, being leached through the soil profile during periods when there is no crop cover, or the plant cover is dormant.

Rates of manure and/or compost application should be determined based on the crop requirement *and* on the elemental content of the material being applied. Estimated elemental content values are not sufficient, but should be based on a laboratory analysis conducted by a laboratory experienced in this type of assay work.

## 20.4   OTHER ORGANIC MATERIALS

Various organic materials, mostly by-products, have use as organic fertilizers for one or more of the essential plant nutrient elements. Their availability varies widely, as well as suitability for general use. Although containing one or more of the essential plant nutrient elements, elemental content may not be sufficient to meet crop requirements; or the form of the element may not make the element readily available, requiring a period of either decomposition and/or solubilization. These materials and their elemental content include:

| Organic Material | Elemental Content |
|---|---|
| Blood meal, dried blood | 15% N, 1.3% P, 0.7% K |
| Dried blood | 12% N, 3.0% P, 0% K |
| Bone meal | 3.0% N, 20.0% P, 0% K, 24 to 30% Ca |

| Composted cow manure | 2% N, 1% P, 1% K |
| Cottonseed meal | 6% N, 2 to 3% P, 2% K |
| Fish emulsion, fish meal | 10% N, 4 to 6% P, 1% K |
| Guano (bat) | 8% N, 4% P, 29% K average, but varies widely, 24 trace minerals |
| Guano (bird) | 13% N, 8% P, 2% K, 11 trace minerals |
| Hoof and horn meal | 14% N, 2% P, 0% K |
| Langbeinite | 0% N, 0% P, 22% K, 22% S, 11% Mg |
| Leatherdust, leather meal | 5.5 to 22% N, 0% P, 0% K |
| Kelp meal, liquid seaweed | 1% N, 0% P, 12% K |

Although an organic material contains an essential plant nutrient element, the availability of that element for root absorption may be quite low, and therefore insufficient to meet the plant's nutrient element requirement for the crop plant(s) being grown. Total content and that existing in a soluble form, available for immediate utilization by growing crop, depends on the substance itself and its biodegradability. Estimated N-P-K content and plant nutrient element release for various organic and natural substances are as follows:

| Material | Estimated N-P-K | Rate of Nutrient Element Release |
|---|---|---|
| **Organic** | | |
| Alfalfa meal | 2.5-0.5-2.0 | Slow |
| Blood meal | 12.5-1.5-0.8 | Medium-fast |
| Bone meal | 4.0-21.0-0.2 | Slow |
| Cottonseed meal | 7.0-2.5-1.5 | Slow-medium |
| Crab meal | 10-0.3-0.1 | Slow |
| Feather meal | 15-0.0-0.0 | Slow |
| Fish meal | 10-5-0.0 | Medium |
| Bat guano | 5.5-8-1.5 | Medium |
| Seabird guano | 12.3-11-2.5 | Medium |
| Kelp meal | 1.0-0.5-8.0 | Slow |
| Dried manure | Depends on source | Medium |
| Soybean meal | 6.5-1.5-2.4 | Slow-medium |
| Worm castings | 1.5-2.5-1.3 | Medium |
| **Natural Inorganic** | | |
| Granite meal | 0.0-0.0-4.5 | Very slow |
| Green sand | 0.0-1.5-5.0 | Very slow |
| Colloidal phosphate | 0.0-16-0.0 | Slow-medium |
| Rock phosphate | 0.0-18-0.0 | Very slow-slow |
| Wood ash | 0.0-1.5-5.0 | Fast |

## 20.5 SOIL AND PLANT FACTORS

Most organic substances, when used as fertilizers, are not well balanced in terms of their essential major plant nutrient element contents. When having a relatively low content for a particular plant nutrient element, a high application rate will be required in order to supply that particular element with the danger that an accompanying plant nutrient element, or other elements, will be applied in excess, thereby creating an imbalance among the essential plant nutrient elements. This situation is not an unusual occurrence and can be the source of a "created" insufficiency condition that can lead to poor plant growth and soil infertility.

# 21 Fertilizer Placement

How and when and in what form a fertilizer is placed in a soil or within the root zone ranks as important when applying the correct fertilizer formulation and quantity.

## 21.1 OBJECTIVES

The three basic fertilizer placement objectives are to

1. Achieve efficient use of essential plant nutrient elements from plant emergence to maturity.
2. Avoid fertilizer-induced salt injury to plants.
3. Prevent or reduce any harmful effects on the environment.

With the exception of foliar fertilization, fertilizers are either

- Broadcast on the soil surface and then cultivated into the soil.
- Applied in bands at specific widths under the soil surface at a particular depth prior to planting.
- Applied alongside the seed row or below the seed row at planting.
- Dissolved in applied irrigation water.

The following are the distinctions between location of placement and different methods of placement needed to understand the various types of fertilizer placement practices:

| Location of Placement | Method of Placement |
|---|---|
| Soil surface | Broadcast |
| | Stripped or dribbled |
| | Sidedressed and topdressed |
| | Dissolved in irrigation water |
| Below soil surface | Broadcast, plowed under |
| | Broadcast, incorporated in varying degrees by other tillage operations |
| | Sidedressing |
| | Row application with seed, pop-up or starter |
| | Banded apart from seed, starter |
| | Deep banding, in fall or spring before seeding, also referred to as dual application |
| | Deep placement, knifing, tillage implement application, pre-plant, banding, etc. |
| In irrigation water | Including essential plant nutrient elements in applied irrigation water |
| Directly into the plant | Applying water containing an essential element(s) directly on plant leaves |

## 21.2 METHODS OF FERTILIZER PLACEMENT

### 21.2.1 BANDING

Banding is the application of a fertilizer as a narrow intact band that reduces the contact between the applied fertilizer and the soil. Applications may be made prior to, during, or after planting. Banding is particularly effective in increasing the availability of an element whose soil test level is low, when early-season stress may occur due to either cool or wet soil conditions that would limit plant root growth, or when soil compaction limits root development into the soil mass, reducing plant root soil–fertilizer contact. Banding reduces soil–fertilizer contact, which can revert an applied fertilizer element into an unavailable form by fixation.

Generally, banding of P-, K-, and micronutrient-containing fertilizers results in more of the applied element being utilized during the first cropping season than would occur if broadcast. Nitrogen applications under reduced tillage systems are particularly adapted to band applications where first-year responses are essential.

### 21.2.2 SURFACE STRIP OR DRIBBLE BANDING

Strip banding or dribbling is a form of band placement that involves the application of solid or fluid fertilizers in bands or strips of varying widths on the soil surface or on the surface of crop residues. Zones of high element concentration are produced, which improves element-use efficiency.

Typically, an applied fertilizer contacts only 25% to 30% of the soil surface. If these surface strip applications are followed by tillage, the concentration effect is diluted to something between broadcast applications and sub-surface banding where the concentrated zones remain intact.

### 21.2.3 DEEP BANDING

The term "deep banding" refers to pre-plant applications of a fertilizer placed 2 to 6 inches below the soil surface. Some applications can be deeper, as much as 15 inches, depending on the soil type, fertility status, and cropping system. The applied fertilizer may be in solid, fluid, or gaseous form, such as anhydrous ammonia ($NH_3$). Concentrated zones of elements are produced, as either streams, pulses, or points, depending on the design of the applicator. In some situations, this fertilization technique is performed many months before the next crop is seeded, often in conjunction with a tillage operation.

In some areas of the coastal plains in the southern United States, deep banding with a shank running immediately ahead and below the planting unit is common. Using this technique, large amounts of fertilizers can be placed immediately below the seed and existing compacted zones are opened for root development.

"Dual application," another term for deep banding, implies simultaneous application of anhydrous ammonia ($NH_3$) as the N source and either fluid or solid P-, K-, and S-containing fertilizers. Otherwise, deep banding terminology can imply the use of either fluid or solid fertilizers.

### 21.2.4 HIGH PRESSURE INJECTION

This method of application of liquid fertilizer is a form of deep banding that employs high pressure to force a constant stream or pulse of liquid fertilizer through the surface residue and into the soil. No physical disturbance of the soil is necessary. Pressures range from about 2,000 to 6,000 psi. Advantages of this method of application would be the same as for deep banding, but the depth of penetration would be less, depending on the soil type and structure.

### 21.2.5 POINT INJECTION OF FLUIDS

This technique of liquid fertilizer application is another variation of deep placement that employs a spoked wheel to physically inject fertilizer at points about 8 inches apart to a soil depth of about 4 to 5 inches. Spacing between the wheels can be varied according to the crop being fertilized. A rotary valve in the wheel hub dispenses fertilizer to the "down" spike using a positive displacement pump.

### 21.2.6 POINT PLACEMENT OF SOLIDS

Deeply placed nests, pockets, clumps, or very large (super granules) pieces of solid fertilizers are in the developmental stage. This technique has been conceived primarily for concentrating ammoniacal sources of N in localized zones at approximately the same soil depths as regular deep banding. Solid sources of conventionally sized urea $[CO(NH_2)_3]$ or ammonium sulfate $[(NH_4)_2SO_4]$ are often used for nesting or super granules of urea are substituted and individually placed.

### 21.2.7 STARTER

Starter fertilizer application is a form of band placement applied at planting either below the seed, or to the side and below the seed. "Pop-up" is another term sometimes used in connection with starter terminology, applying a fertilizer solution, low in concentration in order to avoid germination and seedling damage, on the seed at planting as a means of stimulating early growth. Fertilizer rates in direct seed contact must be kept low to avoid germination and seedling damage.

### 21.2.8 SIDEDRESSING

Sidedressing involves band placement of a fertilizer beside the crop row after plant emergence, either surface banding or sub-surface placement of a fluid or solid fertilizer as well as anhydrous ammonia ($NH_3$). The form and rate depend on the crop requirement. With sub-surface application, distance from the row and depth of placement are important in terms of minimizing root pruning.

### 21.2.9 FERTIGATION

Fertigation is the inclusion of a fertilizer or one or more essential plant nutrient elements dissolved in the applied irrigation water. The advantage is that an essential plant nutrient element is being applied in order to sustain plant growth, ensuring sufficiency over the growth cycle, which would include fruiting and fruit development. Such a practice is used when soil conditions are such that the availability of an element would be reduced if soil was applied prior to or at planting. In addition, with this practice, controlled availability of an essential plant nutrient element can be obtained, which would avoid either "luxury" consumption or periods of insufficiency that might occur at certain critical periods of plant development.

### 21.2.10  FOLIAR FERTILIZATION

Foliar application of an essential plant nutrient element, either as a primary means or as a supplement when the symptoms of a deficiency are observed, will not always ensure sufficiency or correct a deficiency. Foliar application has proven to be a successful method for correcting several micronutrient (B, Mn, Fe, and Zn) deficiencies, and is of questionable value in correcting a major essential element deficiency. Foliar application of any of the essential plant nutrient elements as a boost to ensure adequacy has proven ineffective.

Leaf structures are such that the absorption of a foliar-applied essential plant nutrient element does not always readily occur, including its translocation to other portions of the plant. For example, when an Mn-containing solution is sprayed onto the leaves of a soybean plant showing an Mn deficiency symptom, only on those areas of the leaves where the solution makes contact will the deficiency symptoms disappear.

The absorption of a foliar-applied essential plant nutrient element depends on the composition of the foliar solution, the condition of the plant, the time applied, and the weather conditions at the time of application.

1. The foliar solution composition factors include
   - Form (inorganic or organic matrix) of the essential plant nutrient element being applied
   - Type of wetting agent used
2. The plant conditions include
   - Leaf characteristics [physical size and maturity, presence of pubescence, surface characteristics (waxy, smooth, etc.)]
   - Stage of plant growth (vegetative, approaching maturity, flowering, fruiting)
3. The time factor is time of day.
4. The weather factors include
   - Air temperature
   - Relative humidity
   - Wind

Maximum absorption of an applied plant nutrient element occurs when

- The plant nutrient element is in low concentration.
- Applied solution contains a wetting agent.
- The plant tissue is fully turgid and in the maturing stage of development.
- Applied in the late afternoon or just before dawn.
- The air temperature is cool, has a high relative humidity and there is no wind.

# 22  Soil Water, Irrigation, and Water Quality

This topic cannot be adequately covered in this treatise as it is an extensive topic in and of itself. Therefore, the reader should refer to texts that deal specifically with each topic for greater coverage. The purpose here is to present those water terminology terms related specifically to soil characteristics, basic systems of irrigation, and that define those water characteristics required for irrigation purposes.

## 22.1  SOIL WATER TERMINOLOGY

As with any scientific system, there develop terms that define characteristics that relate to the topic. The following terms are used to define classes of plant-available water, available water, and soil moisture status associated with crop availability.

1. Classes of plant-available water:
   - *Saturation:* Gravitational water, rapid drainage
   - *Field capacity:* Capillary water, slow drainage
   - *Permanent wilting:* Hydroscopic water, essentially no drainage
2. Available water between field capacity and permanent wilting:
   - *Unavailable water:* That below permanent wilting
   - *Field capacity:* The amount of water remaining in the soil after gravitational flow has stopped
   - *Permanent wilting point:* Amount of water remaining in the soil when plants begin to wilt—water is held so tightly that plant roots cannot absorb.
3. Soil moisture status related to plant availability:
   - *Saturation percentage:* 1/10 atmosphere near saturation
   - *Field capacity:* 1/3 atmosphere, 1/2 saturation percentage
   - *Permanent wilting point:* 15 atmospheres, 1/4 saturation percentage

The water content of a saturated soil (i.e., its saturation percentage) is about twice the field capacity and about four times the permanent wilting point. The relationships among saturation percentage, field capacity, and permanent wilting point generally hold consistent for all soils, from clay loams to sandy soils. Consumptive water use is the sum of two factors: transpiration (loss of water as vapor from leaf surfaces) and evaporation (loss of water from the soil surface).

## 22.2 SOIL FACTORS AFFECTING SOIL WATER-HOLDING CAPACITY AND MOVEMENT

Soil factors that affect water movement and availability are texture, structure, and drainage, whether natural or obtained by an installed drainage system. Soil-water characteristics, water-holding capacity, and infiltration rates are based on soil texture.

Approximate soil-water characteristics for typical soil textural classes are

| Characteristic | Sandy Soil | Loam Soil | Clayey Soil |
|---|---|---|---|
| Dry weight, 1 cubic foot | 90 lb | 80 lb | 75 lb |
| Field capacity, % of dry weight | 10% | 20% | 35% |
| Permanent wilting percentage | 5% | 10% | 19% |
| Percent available water | 5% | 10% | 16% |
| Water available to plants: | | | |
| pounds/cubic feet | 4 lb | 8 lb | 12 lb |
| inch/feet depth | ¼ in | 1½ in. | 2¼ in. |
| gallons/cubic feet | ½ gal | 1 gal | 1½ gal |

The approximate amounts of water held by various soil textures are

| Soil Texture | Inches of Water Held/Foot of Soil | Maximum Rate of Irrigation (in./hr) Bare Soil |
|---|---|---|
| Sand | 0.5–0.7 | 0.75 |
| Fine sand | 0.7–0.9 | 0.60 |
| Loamy sand | 0.7–0.11 | 0.50 |
| Loamy fine sand | 0.8–1.2 | 0.45 |
| Sandy loam | 0.8–1.4 | 0.40 |
| Loam | 1.0–1.8 | 0.35 |
| Silt loam | 1.2–1.8 | 0.30 |
| Clay loam | 1.3–2.1 | 0.25 |
| Silty loam | 1.4–2.5 | 0.20 |
| Clay | 1.4–2.4 | 0.15 |

*Source:* From *Western Fertilizer Handbook, Horticultural Edition*, 1990.

The approximate water storage capacities of soils by soil texture are

| Soil Texture | Total Storage (in./ft) | Available Water (in./ft) |
|---|---|---|
| Coarse sand | 1.0–1.5 | 0.6–0.8 |
| Sandy loam (coarse to medium) | 2.0–2.5 | 1.0–1.5 |
| Silts and loams (medium) | 3.5–4.0 | 1.6–2.0 |
| Clay loams (medium to fine) | 4.0–4.5 | 2.0–2.5 |
| Clays (fine) | 4.5–5.0 | 1.6–2.0 |

The infiltration of water into a soil is determined by its textural class. The suggested maximum water infiltration rates for various soil textures are:

| Soil Texture | Infiltration Rate (in./hr) |
|---|---|
| Sand | 2.0 |
| Loamy sand | 1.8 |
| Sandy loam | 1.5 |
| Loam | 1.0 |
| Silt and clay loam | 0.5 |
| Clay | 0.2 |

Infiltration rates are also affected by the interaction between soil texture and the physical condition (structure of the soil:

| | Infiltration Rates | |
|---|---|---|
| Soil Texture | Good Physical Condition (in./hr) | Poor Physical Condition (in. hr) |
| Coarse sand | 2.0–3.0 | 1.2–1.6 |
| Fine sand | 1.8–2.0 | 0.8–1.2 |
| Sandy loams | 1.2–1.8 | 0.5–1.0 |
| Silts and loams | 1.0–1.2 | 0.3–0.4 |
| Clay loams | 0.2–0.5 | 0–0.3 |
| Clays | 0.2–0.5 | 0.02 |

Soil texture is also a factor affecting the percent of plant-available water and depending on soil water tension levels:

| Tension Less Than (bars)[a] | Loamy Sand | Sandy Loam | Loam | Clay |
|---|---|---|---|---|
| | | % Plant-Available Water | | |
| 0.3 | 55 | 35 | 15 | 7 |
| 0.5 | 70 | 55 | 30 | 13 |
| 0.8 | 77 | 63 | 45 | 20 |
| 1.0 | 82 | 68 | 55 | 27 |
| 2.0 | 90 | 78 | 72 | 45 |
| 5.0 | 95 | 88 | 80 | 75 |
| 15.0 | 100 | 100 | 100 | 100 |

[a] 1 bar = 100 kilopascals.

## 22.3  DRAINAGE

Many soils have natural drainage characteristics, allowing water to easily drain from the rooting zone with each rainfall. Soil texture and structure characteristics will determine how much water will infiltrate or run off with each rainfall event. The condition of the subsoil will also determine how easily and quickly water will drain from the surface soil. For some soils, depending on both the surface soil conditions (depth, texture, structure, and organic matter content) and the underlying subsoil (texture and underlying structure), the presence of a clay or plow pan will require some form of an installed drainage to carry away excess moisture. The design and installation of a drainage system require professional skills.

## 22.4   IRRIGATION METHODS

The purposes for irrigation are to

- Add water essential for plant growth to the soil.
- Leach or reduce the concentration of salts in the soil.
- Soften the soil in order to reduce the effects of tillage pans and clods.
- Cool the soil and atmosphere around the plant.
- Reduce the hazard of frost.
- Delay bud formation by evaporative cooling.

There are four methods of crop irrigation—flood (basin and furrow), sprinkler, drip, and trickle—each having its own characteristics in terms of required equipment, and suitability based on crop and soil conditions.

Flood and sprinkler irrigation are the most inefficient in their use of water, while with drip and trickle irrigation, water is placed at or near the base of the plant, wetting only that area of the soil occupied by the majority of plant roots. Manipulation of the underlying water table, requiring control of the normal drainage system by the use of dams and/or blocking devices, has application where such conditions exist.

The placement of water-supplying devices within or just under the root zone is a unique means for supplying water that requires a special water delivery system and is usually installed during soil preparation or at planting. Plant growth and yield responses to each of these irrigation methods have generally shown that best results occur when water is applied when the atmosphere demand is at its maximum. Water use efficiency is related to irrigation method and crop conditions. It is generally conceded that with most irrigation systems, more water is applied than specifically required by a crop. When to irrigate is more difficult to schedule in humid climates, determining when to irrigate when anticipated rainfall events may occur.

Fertigation, the adding of an essential element or elements to applied irrigation water, is primarily applicable to sprinkler, drip, and trickle irrigation systems (see page 173). What essential plant nutrient element or elements to include in applied irrigation water, their concentration and form, are factors that require considerable skill. The danger here, related to need and environmental conditions, occurs when an essential plant nutrient element(s) is needed and water is not. For high-value crops, fertigation is practiced based on monitoring the elemental content of a crop plant by means of plant analyses or tissue tests. Usually such crops are grown under plastic culture, which keeps rainfall out of the rooting bed.

## 22.5 IRRIGATION WATER QUALITY

Water used to irrigate should be free of constituents that could harm plants, including both organic substances and inorganic elements. The inorganic elements or compounds that can be detrimental to plants are Na if greater than 70 ppm, B if greater than 2.0 ppm, and the anions chloride ($Cl^-$) if greater than 100 ppm and bicarbonate ($HCO_3^-$) if greater than 40 ppm. Water that is calcium carbonate ($CaCO_3$) saturated will have a pH of 8.3, requiring acidification if used. High (greater than 70 ppm) Na content water may have a pH greater than 8.3 and not be suitable for use.

The quality of irrigation water, whether well, surface, or rainwater, will be determined by what exists in its surroundings. Underground waters can be relatively pure when drawn from sand or gravel aquifers versus that drawn from limestone or similar strata that will add Ca, Mg, and bicarbonates. Some well waters will also contain Fe and the sulfide anion ($S^-$) anion. Surface waters from rivers, lakes, and ponds will contain substances being washed into the water from the surrounding environment. Even rainwater will contain substances suspended in the atmosphere that are brought down with the rain. Water suspended in the atmosphere is derived from what is being carried into the air by both natural and human activities. Therefore, to ensure the suitability of irrigation water, what is in the water should be determined before use.

The levels of commonly found substances in water that determine its suitability for irrigation use have been fairly well established. The following is one of the interpretation guidelines defining water quality parameters for irrigation use:

| Potential Problem | Degree of Restriction on Use | | | |
|---|---|---|---|---|
| | Units | None | Slight to Moderate | Severe |
| pH | Normal range 6.5 to 8.4 | | | |
| Salinity, $EC_w$ | ds/m | <0.7 | 0.7–3.0 | >3.0 |
| TDS | mg/L | <450 | 450–2,000 | >2,000 |
| Infiltration[a]: | | | | |
| SAR = 0–3 and $EC_w$ | ds/m | >0.7 | 0.7–0.2 | <0.2 |
| SAR = 3–6 and $EC_w$ | ds/m | >1.2 | 1.2–0.3 | <0.3 |
| SAR = 6–12 and $EC_w$ | ds/m | >1.9 | 1.9–0.5 | <0.5 |
| SAR = 12–20 and $EC_w$ | ds/m | >2.9 | 2.9–1.3 | <1.3 |
| SAR = 20–40 and $EC_w$ | ds/m | >5.0 | 5.0–2.9 | <2.9 |
| Specific ion effects: | | | | |
| Sodium[b] | | | | |
| Surface irrigation | SAR | <3 | 3–9 | >9 |
| Sprinkle irrigation | mg/L | <3 | >3 | |
| Chloride[b] | | | | |

| | | | | |
|---|---|---|---|---|
| Surface irrigation | meq/L | <4 | 4–10 | >10 |
| Sprinkle irrigation | meq/L | <3 | >3 | |
| Boron | mg/L | <0.7 | 0.7–3.0 | >3.0 |
| Bicarbonate[c] | meq/L | <1.5 | 1.5–8.5 | >8.5 |

[a] At a given sodium adsorption ratio (SAR), infiltration rate increases as water salinity increases.

[b] For surface irrigation, most tree crops and woody plants are sensitive to sodium and chloride; use values shown.

[c] Applies to overhead sprinkling only.

Irrigation water quality parameters can also be defined for use on specific crops or growing systems, such as that acceptable for use for bedding plants:

| Variable | Plug Production | Finish Late and Pots |
|---|---|---|
| pH[a] (acceptable range) | 5.5 to 7.5 | 5.5 to 7.5 |
| Alkalinity[b] | 1.5 meq/L | 2.0 meq/L |
| | (75 ppm) | (100 ppm) |
| Hardness[c] | 3.0 meq/L | 3.0 meq/L |
| | (150 ppm) | (150 ppm) |
| EC | 1.0 mS | 1.2 mS |
| Ammonium N | 20 ppm | 40 ppm |
| Boron | 0.5 ppm | 0.5 ppm |

[a] pH not very important alone; alkalinity more important.

[b] Moderately higher alkalinity levels are acceptable when lower amounts of limestone incorporated into the substrate during its formulation. High alkalinity levels require acid injection into water source.

[c] High hardness values are not a problem if calcium and magnesium concentrations are adequate and soluble salt level is tolerable.

Source: Knott's Handbook for Vegetable Growers, 4th edition, 1997.

Salinity is another factor that can determine the suitability of irrigation water. The expected crop response to salinity is:

| Salinity (ECe, mmho/cm, or dS/m) | Crop Response |
|---|---|
| 0–2 | Salinity effects mostly negligible |
| 2.4 | Yields of very sensitive crops may be restricted |
| 4–8 | Yields of many crops restricted |
| 8–16 | Only tolerant crops yield satisfactorily |
| >16 | Only a few very tolerant crops yield satisfactorily |

## 22.6  WATER TREATMENT PROCEDURES

Passing water through fine sand or a similar type of filter will remove suspended particles, but will not remove dissolved substances. However, charcoal and

Millipore-type filters will remove organic molecules, the extent of removal depending on the characteristics of the filter material. If irrigation water is surface water or rainwater, such filtering is desirable. Passing an irrigation water source through an ion exchange system or reverse osmosis device ("RO water" means that obtained by reverse osmosis), all of the dissolved substances will be removed. Using an ion exchange procedure, the ion exchanged will remain in the generated water. For example, for Na-based ion exchange devices, Na will be the element remaining in the generated water. Water characteristics and how to treat water for purification is discussed in a 1997 article in *Today's Chemist* (Anonymous, 1997).

## 22.7 WHAT IS WATER?

Biologically, water is referred to as the "universal solvent." It is also unique in that the water molecule is a polar compound carrying both a positive and negative charge. It is the only substance that can exist in all three physical phases—solid, liquid, and gaseous vapor—in the same environment. Recent articles on the characteristics of water have been written by Morgan (2005) and Folds (2009).

# Section VI

Methods of Soilless
Plant Production

# 23 Hydroponics

The technique for growing plants in an enriched aerated water solution had its debut in the mid-1850s when researchers developed this technique as a means for determining those elements that, when absent from a nutrient solution formulation, would result in either the plant dying or its growth significantly impaired. This technique is still in use today by researchers as a means of modifying the elemental content of the solution bathing the roots. The solution used to bathe the plant roots is called a nutrient solution.

## 23.1 HYDROPONICS DEFINED

The word *hydroponics* was formed by combining two Greek words, *hydro* meaning water and *ponos* meaning work; the combination means "working water," as a method for growing plants without soil. Other words have been used to describe the growing of plants without soil, including *aqua (water) culture, nutriculture, hydroculture, soilless culture, tank farming,* and *chemical culture.*

## 23.2 HISTORICAL EVENTS

In the late 1920s, hydroponics was hailed as the future for the large-scale production of food plants, thereby eliminating the need for soil. The economic conditions of that period, followed by World War II, took the focus away from this grand idea. However, during World War II on some islands in the Pacific, the U.S. Army used a hydroponic flood-and-drain system for the production of tomatoes and lettuce, providing these two fresh vegetables for combat troops operating in that area. Following World War II, similar commercial hydroponic farms were established, mainly in semi-tropical and tropical areas. The book by Eastwood (1947) describes these early military and commercial applications. However, production costs, coupled with the inability to control root disease, terminated most of these enterprises. However, the commercialization of hydroponics continues to attract researchers and entrepreneurs, devising ways to grow plants hydroponically that are economically viable and allow control of the rooting environment.

Today, there are a number of different hydroponic growing systems in use adapted for various commercial and hobby uses. For the commercial production of tomatoes, cucumbers, and peppers, drip irrigation hydroponics is the preferred method; NFT (nutrient film technique) for the production of lettuce, strawberries, and herbs; and a standing aerated nutrient solution method for the production of lettuce and herbs. Hydroponic growing is primarily practiced in greenhouses or controlled environment chambers, and only to a limited degree in the open environment. The books by Jones (2005, 2011) and Resh (2001) are major texts on this subject.

## 23.3  HYDROPONIC TECHNIQUES

There are two hydroponic techniques defined by the way a nutrient solution is used:

1. An open system in which the nutrient solution is passed just one time through the plant root mass or rooting medium, and then discarded.
2. A closed system in which the nutrient solution, after passing through the plant root mass or rooting medium, is recovered and recirculated.

Either technique can be applied to all hydroponic growing systems.

The open system can be wasteful in its use of water and reagents, depending on how either the root mass or rooting medium pass-through is conducted. The nutrient solution can be formulated by injecting elemental concentrates into a flowing stream of water, therefore not requiring a large storage vessel for the formulated nutrient solution. This system has the advantage that plant roots are being exposed to a constant-composition nutrient solution. The nutrient solution, after passing either through the plant root mass or rooting medium, can be considered a hazardous waste, therefore requiring specialized recovery and disposal procedures.

For the closed system, there must be a means of recovering the nutrient solution after it passes through either the plant root mass or rooting medium. Therefore, nutrient solution storage vessels are required, their number and volume capacity depending on the size of the hydroponic growing system. There are options on how the recovered nutrient solution is treated before its recirculation. Water is normally added to bring the nutrient solution back to its original volume. It then can be either recirculated without treatment, or its pH and nutrient element contents determined and adjusted to bring both back to their original levels, and/or filtered to remove suspended materials, and/or sterilized by heat, ozone injection or ultraviolet (UV) exposure, and/or aerated by bubbling air through it. Without these treatments, there is the possibility for the occurrence of plant nutrient element insufficiencies as well as plant root disease. Depending on how the recovered nutrient solution is treated, it is possible to continuously use it over the entire life cycle of the plants being grown.

## 23.4  HYDROPONIC GROWING SYSTEMS

Today, there are basically six hydroponic growing systems in use, three systems in which the plants are grown without the use of a rooting medium, the remaining three with plants rooted in either an inorganic, organic, or an inorganic/organic mixture. Each hydroponic growing system has its own unique characteristics and requirements, having specific applications in terms of plant characteristics and number of plants per area and/or volume of nutrient solution. Depending on the growing system and its size, some can be manually operated, while others require electrical power, pumps, timers, and specialized equipment in order to deliver the nutrient solution to or through the plant root mass or rooting medium.

### 23.4.1   Systems Without the Use of a Rooting Medium

### 23.4.1.1   Aerated Standing or Circulated Aerated Nutrient Solution

Plant roots are suspended in a depth of nutrient solution that is continuously aerated, the size of the rooting vessel and nutrient solution composition and volume depending on the plant species being grown. A commercial application is the growing of short stem plants, such as lettuce and herbs, with the plants being placed in openings in a raft that is floated on the surface of a depth (usually 5 to 6 inches) of nutrient solution that is being continuously aerated and/or circulated as the means for aeration. By adjusting the depth of nutrient solution, the effects due to changing environmental conditions are minimized, and with increasing volume, the less frequently the nutrient solution will require modification and/or replacement. An objective would be to have a sufficient volume of nutrient solution so that a crop can be carried from the seedling stage to harvest without the need for either modification or replacement.

There is no set number of plants, volume of nutrient solution, or elemental content/concentration parameters for this hydroponic method. In general, the greater the volume of nutrient solution per plant, the lower the elemental concentration and the less frequently the nutrient solution will have to be renewed or replaced with fresh solution. Plant species, rate of growth, and stage of growth are also factors that can affect these determinations. Plants will grow best when the volume of nutrient solution per plant is large, with the nutrient solution elemental concentration lower than that recommended for most nutrient solution formulations.

*Advantages*: The standing-aerated nutrient solution system is easy to assemble and operate. For the floating raft system, the nutrient solution-holding pool can be constructed using plastic sheeting as the liner of varying dimensions and depth, with minimum requirements for pumps, piping, and storage containers; a child's plastic swimming pool can be used by the home gardener as the nutrient solution container; the plastic used to form the nutrient solution pool can be either discarded after use or cleaned and sterilized for the next crop; for the raft, Styrofoam is a suitable, inexpensive and durable substance that can be either discarded when the crop is harvested, or reused after sterilization; the system can be used outdoors when rainfall events are infrequent and/or of low amounts for any one event.

*Disadvantages*: Normally, the standing-aerated nutrient solution system is for the growing of a single plant per vessel of sufficient size to accommodate the plant roots, requiring constant aeration and periodic replacement of the nutrient solution, depending on the plant being grown and its growth rate. For the raft growing system, it is only suited for the production of short stem plants and those that come to maturity in a short time period (30 to 45 days); the nutrient solution bed must be level and water-tight; the nutrient solution requires constant aeration and/or circulation to maintain the desired oxygen ($O_2$) level in solution; root disease is a constant threat; temperature control of the nutrient solution can pose a problem in environments that have widely varying air temperatures; method can be wasteful of water and reagents when the nutrient solution is eventually discarded.

### 23.4.1.2   Nutrient Film Technique (NFT)

In 1975, the Nutrient Film (sometimes the word "Flow" is used instead of "Film") Technique was introduced, its identification being NFT, the acronym coined by its inventor, Allen Cooper (1976). With its introduction came increased interest in the commercial production of tomato, cucumber, pepper, and lettuce where plant roots are suspended in a flow of nutrient solution down an enclosed sloping (2% to 3% slope) trough. Initially, NFT was considered a major advancement in hydroponics; however, when put into commercial use, operational flaws have limited its application primarily to the production of lettuce.

The number of plants, volume of nutrient solution, its elemental concentration, and frequency of application are the significant parameters when using this hydroponic method. Plant species, rate of growth, and stage of growth are also factors that will determine plant nutritional sufficiency. Normally, the frequency of nutrient solution application is based on the water needs of the plant, which can vary with the number of plants, their stage of growth, and atmospheric demand. The most common error with this hydroponic growing method is using a nutrient solution whose elemental concentration is greater than what the plant needs, thus resulting in elemental accumulation in the root mass and the possible occurrence of plant nutrient element insufficiencies. Alternating between a nutrient solution application and water-only can be used to ensure that sufficient water is supplied to meet the atmospheric demand of the plants, while minimizing the potential of a nutrient element insufficiency due to excess.

*Advantages*: Easy and inexpensive to set up using a plastic sheet constructed trough that can be discarded after use; well suited for the growing of lettuce in a variety of trough designs.

*Disadvantages*: The system requires nutrient solution storage vessels, their number and volume capacity depending on the size of the NFT system; the length of the trough will determine plant performance at the end of the trough due to changes in oxygen ($O_2$) and elemental content of the flowing nutrient solution; the system is not suited for long-term crops (such as tomato, cucumber, and pepper), for as plant roots fill the trough, the flow of nutrient solution down the trough is impeded, anaerobic conditions developing in the root mass, resulting in root death; recirculation of the nutrient solution will require volume, pH, and essential element adjustments as well as filtering to remove suspended materials and sterilization.

### 23.4.1.3   Aeroponics

Plant roots are suspended in an enclosed chamber and intermittently bathed with a spray or mist of nutrient solution. The roots are essentially growing in air that promotes both root generation and extensive growth; therefore aeroponics can be used for the propagation of plant species that are difficult to root from woody stem cuttings. The finer the water droplets, the greater will be the potential for adherence of water on the roots. In some system designs, a shallow pool of nutrient solution is allowed to accumulate in the bottom of the rooting chamber so that a portion of the root system can lie in this pool to have access to water sufficient to meet the demand when plants are grown under high atmospheric demand conditions.

Because very little of the applied nutrient solution adheres to the root surface, an elementally concentrated nutrient solution may be required to meet the nutritional requirements of the plant.

*Advantages*: Easy and inexpensive to construct the root chamber; procedure well suited for the growing of small plants, such as lettuce and herbs, particularly for those herbs whose roots are the primarily selected plant part.

*Disadvantages*: The nutrient solution needs to be free of suspended materials that can clog the nozzle openings; the size of the nozzle openings will determine droplet size—the smaller the openings, the greater the pump pressure needed to discharge the nutrient solution; under high atmospheric demand conditions, this system of water supply may not be sufficient to meet the plant's water needs; there is the danger of root disease due to the high atmospheric humidity in the rooting chamber; primarily limited for the growing of herbs and lettuce.

### 23.4.2 Systems With the Use of a Rooting Medium

#### 23.4.2.1 Flood-and-Drain (Ebb-and-Flow)

With the commercialization of hydroponic growing beginning in the late 1930s, this was the method used for growing vegetable crops such as tomato, cucumber, pepper, and lettuce. The number and size of nutrient solution storage vessels depend on the number and size of the rooting containers as well as the design characteristics of the entire growing system. The nutrient solution is pumped into the base of the rooting vessel, and then allowed to flow back into the storage vessel by gravity. The rooting medium, either gravel or coarse sand, is placed in a water-tight container that, for short time periods, is periodically flooded with a nutrient solution, the flooding schedule based on the water needs of the plant. The size of the rooting vessel can be considerable, therefore accommodating many plants. This method of hydroponic growing is well suited for use by the home gardener due to its relatively simple design and ease of operation.

The number of plants per area of rooting medium, volume of nutrient solution, its elemental concentration, and frequency of application are significant parameters when using this hydroponic method. Plant species, rate of growth, and stage of growth are also factors that will determine plant nutritional sufficiency. Normally, the frequency of nutrient solution application is based on the water needs of the plant, which can vary with the number of plants, their stage of growth, and atmospheric demand. The most common error with this hydroponic growing method is using a nutrient solution formulation whose elemental concentration is greater than what the plant needs, thus resulting in elemental accumulation as the retained nutrient solution and the formation of precipitates that will eventually lead to possible plant nutrient element insufficiencies. To minimize the accumulation of unused nutrient elements, the provision to periodically circulate water through the rooting medium is recommended, or following a routine of alternating irrigations of water only and nutrient solution.

*Advantages*: The rooting bed can vary in size and type of construction materials; the system is relatively easy to operate; small home-type systems can be operated without electrical power using gravity to flood and drain the rooting bed.

*Disadvantages*: Clean, element- and colloidal-free gravel or coarse sand is difficult to find, and if prepared by the grower, requires the use of strong acids and acid-resistant washing facilities; large rooting bed systems require high-capacity pumps to quickly move the nutrient solution from the storage tank in order to flood the rooting bed; the timing for each irrigation requires a determination of the water needs of the plants; in general, the system is wasteful of water and essential plant elements when the nutrient solution is discarded as well as disposal requirements when considered a hazardous waste; with recirculation of the nutrient solution, volume adjustment is required, its pH and level of essential elements monitored to determine if adjustment is necessary to maintain its initial composition in order to avoid plant nutrient element insufficiencies; filtering to remove suspended materials and sterilization to prevent root disease occurrence is required; the accumulation of essential element precipitates in the rooting medium will begin to impact the nutritional status of the plants, precipitates that cannot be removed by leaching the rooting medium with water; if the rotting medium is to be reused, removing the accumulated precipitates will require acid washing and steam sterilization.

### 23.4.2.2 Drip Irrigation

This is the most universally used commercial hydroponic method for growing tomato, cucumber, and pepper, with the delivery of a nutrient solution to the base of the plant rooted in either bags or pots containing either an inorganic (gravel, sand, perlite, rockwool, volcanic rook) or organic (peat, pinebark, core) medium, or in slabs of either rockwool or coir. Rockwool slabs are currently the rooting medium of choice, with core slabs replacing rockwool slabs where disposal of used slabs is an issue. The nutrient solution is usually formulated with pumps that inject element-containing concentrates into a flow of water. The characteristics of the drip delivery system will be determined by the number of discharge points as well as the size and flow rates of drippers.

Another form of this method has plants placed in the corners of buckets, stacked together at 90-degree angles from each other, forming a tower; the rooting medium is a variety of substances, perlite being most commonly used. A nutrient solution is introduced at the top of the tower, flowing down through the rooting medium by gravity. The nutrient solution is either allowed to flow out of the tower for disposal, or is collected and pumped back to the top of the tower. Depending on the design (stacked buckets or columns with access pockets), the height, and the size of the plant pockets (openings) in the tower, various plant species can be grown, such as flowers, ornamental plants, lettuce, herbs, strawberries, and even tomato, pepper, and vine crops.

For both systems, the flow rate and volume of the nutrient solution applied will depend on the water demand of the plants.

The volume of nutrient solution per plant, its elemental concentration, and frequency of application are the affecting parameters when using this hydroponic method. Plant species, rate of growth, and stage of growth are also factors that will determine plant nutritional sufficiency. Normally, the frequency of the nutrient solution application is based on the water needs of the plant, which can vary with the number of plants per area of rooting medium, their stage of growth, and atmospheric demand.

The most common error with this hydroponic growing method is using a nutrient solution whose elemental concentration is greater than what the plant needs, resulting in elemental accumulation in the retained nutrient solution in the rooting medium. Contributing to this accumulation is the volume of nutrient solution applied with each application, which affects the distribution of the applied elements within the rooting medium. A common recommendation is to apply a sufficient volume of nutrient solution so that there is a slight outflow from the openings in bottom of the rooting vessel. The retained nutrient solution in the rooting medium requires monitoring for its electrical conductivity (EC), so when reaching a certain level, the rooting medium requires leaching with water. If the EC of the retained nutrient solution exceeds a certain point, plant nutritional disorders will occur and plants will begin to wilt when the atmospheric demand is high. By alternating between water and nutrient solution applications, the buildup of retained nutrient solution can be moderated. With element retention in the rooting medium, precipitates of calcium phosphate and sulfate begin to form, also co-precipitating with other elements. When in contact with plant roots, the elements in these precipitates will be brought into solution and absorbed. At this point, the plant has three sources of plant nutrient elements: those in the nutrient solution being applied, those retained in the rooting medium, and those existing as a precipitate in the rooting medium resulting in a loss of plant nutrition control.

*Advantages*: This technique has proven reliable in terms of plant performance, is suited for use with large growing systems, and is fairly efficient in its use of water and plant nutrient elements even when operated as an "open" hydroponic system. Towers can be constructed using buckets and plastic pipe configurations; towers conserve space and allow for large numbers of plants to be grown in a confined space.

*Disadvantages*: Drippers require constant monitoring to ensure that there is a consistent flow of nutrient solution; the injection pumps require periodic inspection and adjustment; the timing schedule and amount of nutrient solution delivered can significantly affect plant growth, therefore requiring skill and experience to determine these factors; accumulation of retained nutrient solution and eventually elemental precipitates in the rooting medium will begin to impact the nutritional status of the growing plants; the EC of the retained nutrient solution requires monitoring, with water-leaching required when the EC exceeds a certain level; water leaching is costly in terms of the water required and collection of the leachate for proper disposal; the rooting medium is normally discarded after one use. For tower systems, the nutrient solution changes in elemental and oxygen ($O_2$) contents as its flows down the column, with the plants at the base of the tower growing slower, and possibly being under both oxygen ($O_2$) and elemental stress; the towers will require periodic rotation so that all plants receive equal sunlight exposure.

### 23.4.2.3   Sub-irrigation

This method of hydroponic growing is a total consumption system as all the applied water as well as the applied plant nutrient elements will pass through the plant;

therefore, elemental concentrations in the nutrient solution must be less (1/2 to 1/3) than that recommended for other hydroponic growing systems in order to avoid insufficiencies due to high accumulation of elements in the plant.

The nutrient solution is delivered to the rooting medium at its base, with a constant level of nutrient solution in the bottom of the rooting vessel maintained either by hand, adding nutrient solution when called for by the position of a float indicator, or by means of a float valve, controlling the flow of nutrient solution from a reservoir. The rooting medium must have wicking ability sufficient to remain moist to a height of at least 4 or 5 inches. The rooting medium selected may contain sufficient nutrient elements so that only water is required. With this method of growing, all of the water and nutrient elements are utilized by the plant.

*Advantages*: Easy and inexpensive to set up, suitable for home garden use; employing a float value system of water and nutrient solution delivery, minimum attention is required; there is no water or nutrient solution effluent to discard; no buildup of accumulating nutrient elements in the rooting medium.

*Disadvantages*: Hand-operating systems require constant monitoring of the position of the float valve and adding water or nutrient solution to maintain a constant depth of nutrient solution in the base of the rooting vessel.

## 23.5   ROOTING MEDIA

The three hydroponic growing systems—flood-and-drain, drip irrigation, and sub-irrigation—require the use of a rooting medium; a list of the inorganic substances is given in Table 23.1 and organic substances in Table 23.2 as given by Morgan (2003). Some of the primary properties of these substances are their inertness, biological characteristics, water-holding capacity, ease of drainage, aeration properties (pore space), volume weight, cation exchange capacity, buffer capacity, and intercellular structure. In some instances, mixtures of these substances are used to obtain certain physical and chemical properties that will give a particular desired volume weight, add to the ease of drainage, improve air porosity, add cation exchange capacity, and increase the water-holding capacity. In addition to their physiochemical properties, some of these substances can be sources of essential elements or potentially toxic substances; therefore their selection as the rooting medium, or in a rooting medium mix, can impact plant growth.

## 23.6   WATER QUALITY

Both organic and inorganic constituents can be found in natural waters (well, ground, river, pond, rainwater) and domestic water supplies, constituents that can significantly affect plants. Surface waters from streams, ponds, and lakes can contain substances than can adversely affect plants. Surface water runoff from agricultural lands can contain residues of substances being applied to the soil (fertilizers and herbicides) and crop plants (pesticides). Rainwater will contain substances suspended in the atmosphere as well as substances that have been deposited on the collecting surfaces. Therefore, those substances deemed undesirable must be removed from water for either irrigation use or for making a nutrient solution.

**TABLE 23.1**

**Characteristics of Inorganic Hydroponic Substrates**

| Substrate | Characteristics |
|---|---|
| Rockwool | Clean, nontoxic (can cause skin irritation), sterile, lightweight when dry, reusable, high water-holding capacity (80%), good aeration (17% air holding), no cation exchange capacity or buffering capacity, provides ideal root environment for seed germination and long-term plant growth. |
| Vermiculite | Porous, spongelike, sterile material, lightweight, high water absorption capacity (five times its own weight), easily becomes waterlogged, relatively high water-holding capacity. |
| Perlite | Siliceous, sterile, sponglike, very light, free-draining, no cation exchange capacity or buffer capacity, good germination medium when mixed with vermiculite, dust can cause respiratory irritation. Perlite is now available that has a basic N, P, K nutrient element charge, sufficient to meet the initial NPK plant needs. |
| Pea gravel | Particle size ranges from 5 to 15 mm in diameter, free draining. Low water-holding capacity, high weight density, may require leaching and sterilization before use. |
| Sand | Varying grain size (ideal between 0.6 to 2.5 mm in diameter), may be contaminated with clay and silt particles that must be removed, low water-holding capacity, high weight density, added to some organic mixes to add weight and improve drainage. |
| Expanded clay | Sterile, inert, range in pebble sizes to 1 and 18 mm, free draining, physical structure can allow for accumulation of water and nutrient elements, reusable if sterilized. |
| Pumice | Siliceous material of volcanic origin, inert, has higher water-holding water capacity than sand, high air-filled porosity. |
| Scoria | Porous, volcanic rock, fine grades used in germination mixes, lighter and tends to hold more water than sand. |
| Polyurethane grow slabs | New material that holds 75% to 80% air space and 35% water-holding capacity. |

The most common recommendation is to use pure water. However, some substance(s) found in water may have no significant effect on plant growth, therefore not justifying the cost for their removal. There may be elements in a water source that could be beneficial to plants, particularly Ca and Mg, common constituents in what is called "hard water." Water quality parameters have been established, setting content levels for varying use suitability standards that would fit all potential uses. The characteristics and maximum elemental content levels in water defining suitability for irrigating plants and/or making a nutrient solution are listed in Table 23.3

Reverse osmosis (RO) effectively removes all dissolved substances found in water as well as small molecules. Passing water through an activated carbon filter will remove organic but not inorganic constituents. Ion exchange methods will remove any dissolved inorganic constituents, but the ion used for the exchange will be found in the generated water.

**TABLE 23.2**

**Characteristics of Organic Substrates**

| Substrate | Characteristics |
|---|---|
| Coconut fiber | Made into fine (for seed germination) and fiber forms (coco peat, palm peat, and coir); useful in capillary systems; high ability to hold water and nutrients; can be mixed with perlite to form medium that has varying water-holding capabilities; products can vary in particle size and possible Na contamination. |
| Peat | Used in seed raising mixes and potting media; can become water-logged and is normally mixed with other materials to obtain varying physical and chemical properties. |
| Composed bark | Used in potting media as a substrate for peat; available in various particle sizes; must be composted to reduce toxic materials in original pinebark (from *Pinus radiata*); high in Mn and can affect the N status of plants when initially used; will prevent the development of root disease. |
| Sawdust | Fresh, uncomposted sawdust of medium to coarse texture good for short-term uses; has reasonable water-holding capacity and aeration; easily decomposed, which poses problems for long-term use; sources of sawdust can significantly affect its acceptability. |
| Rice hulls | Lesser known and used; has properties similar to perlite; free-draining, low to moderate water-holding capacity; depending on source, can contain residue chemicals; may require sterilization before use. |
| Sphagnum moss | Common ingredient in many types of soilless media; varies considerably in physical and chemical properties, depending on origin; excellent medium for seed germination use in net pots for NFT applications; high water-holding capacity and can be easily water-logged; provides some degree of root disease control. |
| Vermicast and compost | Used for organic hydroponic systems, varying considerably in chemical composition and contribution to the nutrient element requirement of plants; can become water-logged; best mixed with other organically derived materials or coarse sand, pomice, or scoria after physical characteristics. |

## 23.7   THE NUTRIENT SOLUTION

### 23.7.1   ELEMENTAL CONTENT

A complete nutrient solution will contain the thirteen essential mineral plant nutrient elements. The major elements are, N, P, K, Ca, Mg, and S, all required in relatively high concentration; and the micronutrients, B, Cl, Cu, Fe, Mn, Mo, and Zn, elements required at lower concentrations.

### 23.7.2   ELEMENTAL FORMS

The thirteen essential mineral plant nutrient elements must exist as ions in solution in order to be absorbed by plant roots, although there is evidence that small molecules such as chelates, proteins, and the boric acid ($H_3BO_3$) molecule can be

## TABLE 23.3
## Characteristics and Elemental Content of Water Suitable for Use when Irrigating Plants and Making a Nutrient Solution

### Characteristics:
pH: 5.0–7.0; Electrical Conductivity (EC): <1 mmho/cm;
Alkalinity (CaCO$_3$/L): 100 ppm, 2 meq/L

### Elemental Content (ppm)

| Major Cations: | | Major Anions: | |
|---|---|---|---|
| Calcium (Ca) | <120 | Nitrate (NO$_3$-N) | <2 |
| Magnesium (Mg) | <24 | Chloride (Cl) | <20 |
| Potassium (K) | <10 | Fluoride (F) | <0.75 |
| Sodium (Na) | <50 | Sulfate (SO$_4$) | <90 |
| Ammonium (NH$_4$) | <8 | Phosphate (PO$_4$) | <3 |
| | | Bicarbonate (HCO$_3$) | <122 |

| Trace Elements: | |
|---|---|
| Aluminum (Al) | <5 |
| Iron (Fe) | <4 |
| Manganese (Mn) | <1 |
| Zinc (Zn) | <0.3 |
| Copper (Cu) | <0.2 |
| Boron (B) | <0.05 |
| Molybdenum (Mo) | <0.001 |

transported through the root membrane. The ionic forms for the major elements and micronutrients are

| Major Elements | Micronutrients |
|---|---|
| Nitrogen: | Boron (BO$_3$$^{3-}$) |
| Ammonium (NH$_4$$^+$) | Chloride (Cl$^-$) |
| Nitrate (NO$_3$$^-$) | Copper (Cu$^{2+}$) |
| Phosphate, tri- (PO$_4$$^{3-}$): | Iron (ferrous, ferric) (Fe$^{2+}$, Fe$^{3+}$) |
| Dihydrogen phosphate (H$_2$PO$_4$$^-$) | Manganese (Mn$^{2+}$) |
| Monohydrogen phosphate (HPO$_4$$^{2-}$) | Molybdenum (MoO$^{3-}$) |
| Potassium (K$^+$) | Zinc (Zn$^{2+}$) |
| Calcium (Ca$^{2+}$) | |
| Magnesium (Mg$^{2+}$) | |
| Sulfate (SO$_4$$^{2-}$) | |

Which form of phosphorus exists in the nutrient solution will depend on pH, dihydrogen phosphate (H$_2$PO$_4$–) and monohydrogen phosphate (HPO$_4$$^{2-}$) in acid solutions and triphosphate ($PO_4^{3-}$) in nutrient solutions when the pH is approaching alkalinity (pH > 7.0).

The terminology that identified the name of an essential element and its ionic form in solution is not consistent for all the essential elements. For the elements K, Ca, Mg. Cu, Mn, and Zn, the same word is used to define both; but for the other essential elements, the elemental word and that in solution in ionic form are different. For N, the two ionic forms are either the nitrate ($NO_3^-$) anion or the ammonium cation ($NH_4^+$); for P, either the dihydrogen phosphate ($H_2PO_4^-$) or monohydrogen phosphate ($HPO_4^{2-}$) anion; for S, the sulfate anion ($SO_4^{2-}$); for B, the borate anion ($BO_3^{3-}$); for Cl, the chloride anion ($Cl^-$); for Fe, either the ferrous ($Fe^{2+}$) or ferric ($Fe^{3+}$) cation; and Mo, the molybdate anion ($MoO_4^{2-}$).

Two ions, the nitrate anion ($NO_3^-$) and potassium cation ($K^+$), present in most nutrient solutions in relatively high concentrations are readily absorbed by the plant root, thereby maintaining the ionic balance within the plant, while all other ions in the nutrient solution are selectively absorbed.

### 23.7.3 CONCENTRATION RANGES AND RATIOS

All thirteen essential mineral plant nutrient elements must be within a particular concentration range as well as in a certain ratio among the elements in a nutrient solution in order to ensure plant nutrient sufficiency with its use. Most nutrient solution formulations contain higher concentrations of some elements than required by the plant, P and N as the nitrate-N ($NO_3$-N) anion being two such elements. In some nutrient solution formulations, the ratios among the major cation elements $K^+$, $Ca^{2+}$, and $Mg^{2+}$ are frequently out-of-balance.

Elements in a nutrient solution interact among themselves, exhibiting both antagonistic as well as synergistic characteristics. For example, among the major cations $K^+$, $Ca^{2+}$, and $Mg^{2+}$, the least competitive is $Mg^{2+}$; therefore its deficiency is likely to occur with the use of some nutrient solution formulations having high concentrations of K and/or Ca when growing Mg-sensitive plants such as tomato. Ammonium-N is also a strong competitive cation ($NH_4^+$) and can reduce the uptake of both $Ca^{2+}$ and $Mg^{2+}$, resulting in a deficiency of either element for those plant species sensitive to either Ca or Mg, or both. For example, a high concentration of $NH_4$-N in a nutrient solution can result in a high incidence of blossom-end rot in tomato and pepper fruits. There exists a synergistic relationship between nitrate-nitrogen ($NO_3$-N) and K, the presence of high concentrations of $NO_3$-N enhancing the uptake of K. The presence of a low concentration of ammonium-nitrogen ($NH_4$-N) in a nutrient solution will enhance the uptake of $NO_3$-N.

### 23.7.4 NITRATE AND AMMONIUM

There is considerable research that indicates that the form of N supplied to the plant can have a significant effect on vegetative growth and fruit yield, as well as association with both plant and fruit quality. A mixture of $NH_4$ and $NO_3$ frequently results in better plant growth, if $NH_4$ does not exceed 25%, as compared to when $NO_3$ is the only N source. For some crops, such as tomato, $NH_4$ in the nutrient solution can increase the incidence of blossom-end rot (BER), a commonly occurring fruit disorder. Therefore, some recommend that $NH_4$ be included in the nutrient solution during the tomato plant's vegetative growth period, but not when the plants are setting and producing fruit.

The question is: should ammonium-N ($NH_4$-N) be included in a nutrient formulation, and if so, at what concentration or ratio? It is recommended that at least 5% to 10% of the total N in the formulation be in the $NH_4$ form, even when growing tomatoes.

### 23.7.5 BENEFICIAL ELEMENTS

There are elements that have been identified as being potentially beneficial to plants but do not meet the established criteria for essentiality, elements that may enhance plant growth under certain circumstances (see Chapter 13, "Elements Considered Beneficial to Plants"). Therefore, should these elements be included in a nutrient solution formulation? The early hydroponic researchers devised an A–Z micronutrient solution as a means of ensuring that potentially influencing elements at trace levels be included in a nutrient solution formulation. Two elements whose suggested essentiality has been supported by recent research are Ni and Si. Nickel is not generally recommended for inclusion in a nutrient solution formulation because its function primarily relates to seed viability. On the other hand, some recommend Si inclusion in a nutrient solution formulation because it provides some degree of leaf disease protection and stalk strength. Silicic acid ($H_4SiO_4$) at 100 ppm is the common recommended level in a nutrient solution formulation. Potassium silicate and sodium silicate are equally suitable sources of Si for addition to a nutrient solution.

Because many of these so-called beneficial elements maybe found as contaminants in either the rooting medium or some of the major element source reagents (see Table 23.4), such as calcium nitrate, potassium nitrate, and magnesium sulfate, there may not be the need to purposely add a mix of beneficial elements. This would also suggest that selecting high-purity reagents may not be the best choice. In addition, the rooting medium itself may contain trace levels of some of these same elements.

### 23.7.6 CHELATES

The use of chelated forms for the micronutrients Fe, Cu, Mn, and Zn in a nutrient solution formulation is questionable. Although chelated forms for these micronutrients have proven to be of value based on certain soil conditions, particularly in alkaline and organic soils and organic soilless rooting media, their use in hydroponic nutrient solutions is not justified in terms of improved availability. When adding a chelate to a mix of elements in solution, the stability of the initial chelate will depend on the concentration of the other ions in solution as well as the pH, which in turn can significantly reduce the chelate effect, therefore eliminating the reason that the chelated form of the element was selected over other elemental forms.

**TABLE 23.4**
**Commonly Used Reagents for Making a Nutrient Solution**

| Major Element Reagents | Formula | Elemental Content (%) |
| --- | --- | --- |
| Ammonium dihydrogen phosphate | $NH_4H_2PO_4$ | N (12), P (28) |
| Calcium nitrate | $Ca(NO_3)_2 \cdot 4H_2O$ | Ca (19), N (15) |
| Magnesium sulfate | $MgSO_4 \cdot 7H_2O$ | Mg (10), S (23) |
| Potassium dihydrogen phosphate | $KH_2PO_4$ | P (32), K (30) |
| Potassium nitrate | $KNO_3$ | K (36), N (13) |

| Micronutrient Reagents | Formula | Elemental Content (%) |
| --- | --- | --- |
| Ammonium molybdate | $(NH_4)_6MoO_{24} \cdot 4H_2O$ | Mo (8) |
| Borax | $Na_2B_4O_{24} \cdot 10H_2O$ | B (11) |
| Boric acid | $H_3BO_3$ | B (16) |
| Copper sulfate | $CuSO_4 \cdot 5H_2O$ | Cu (25) |
| Iron (ferrous) sulfate | $FeSO_4 \cdot 7H_2O$ | Fe (20.1) |
| Iron (ferric) sulfate | $Fe_2(SO_4)_3$ | Fe (14) |
| Iron (ferric) chloride | $FeCl_3 \cdot 6H_2O$ | Fe (20.7) |
| Iron (ferrous) ammonium sulfate | $FeSO_4(NH_4)_2SO_4 \cdot 6H_2O$ | Fe (7) |
| Manganese sulfate | $MnSO_4 \cdot 4H_2O$ | Mn (24) |
| Manganese chloride | $MnCl_2 \cdot 4H_2O$ | Mn (28) |
| Zinc sulfate | $ZnSO_4 \cdot 5H_2O$ | Zn (22) |

It also has been demonstrated that the chelate ethylenediaminetetraacetic acid (EDTA) can be toxic to plants, and therefore some formulations use the chelate diaminetriaminetetraacetic acid (DTPA), which has not been found to be plant toxic.

The chelated form of Fe is commonly selected for use in some nutrient solution formulations while several inorganic forms of Fe, such as

- Iron (ferrous) sulfate ($FeSO_4 \cdot 7H_2O$)
- Iron (ferric) sulfate [$Fe_2(SO_4)_3$]
- Iron (ferric) chloride ($FeCl_3 \cdot 6H_2O$)
- Iron ammonium sulfate [$(NH_4)_2SO_4 \cdot FeSO_4 \cdot 6H_2O$]

will keep Fe in solution, and therefore able to meet the Fe requirement of plants.

### 23.7.7 Nutrient Solution/Water Temperature

The temperature of both the nutrient solution and the water applied to plant roots can significantly affect the plant, when cold (less than the ambient air temperature around the plants) resulting in plant wilting, and when hot (above 90°F) interfering with normal root metabolism. In addition, with increasing temperature, the dissolved oxygen ($O_2$) concentration in water and nutrient solution declines, a factor that can affect plant growth due to inadequate oxygen ($O_2$) surrounding the respiring plant roots.

## 23.7.8   pH and Electrical Conductivity (EC)

It is essential that the pH of the nutrient solution as well as the rooting medium be maintained between pH 5.0 to 6.0. However, it is not necessary to adjust the pH unless the nutrient solution and/or rooting medium is greater than 6.8, or even becomes alkaline (greater than 7.0). Plants can grow quite well at a pH level as low as 4.5. Therefore, no adjustment is generally needed when both the nutrient solution and rooting medium are acidic.

The accumulation of ions, frequently referred to as "salts," in the rooting media is an important parameter. As the EC increases, the ability of plant roots to take up water and nutrient elements from a nutrient solution decreases. Growers are advised to monitor the runoff from the rooting media or extract solution retained in the media for its EC; and when exceeding a certain level, the rooting is to be leached with water. An increasing EC in the rooting media suggests that the elemental concentration in an applied nutrient solution is either too high or the frequency of application is greater than needed.

One procedure that can keep the EC from increasing in the rooting medium is to irrigate with a nutrient solution, followed by a water-only irrigation, the scheduling of both based on that needed to satisfy the water requirement of the plant. Another procedure is to apply a nutrient solution at the beginning of the day, to be followed by water-only the rest of the day, its scheduled application sufficient to meet the water requirements of the plant.

## 23.7.9   Other Factors

Filtering the nutrient solution to remove suspended particles is necessary when using the aeroponic (page 190) and drip irrigation (page 192) growing systems in order to prevent the clogging of nozzles and drippers, respectively.

The temperature of the applied nutrient solution can significantly affect plant growth. The rule of thumb is to maintain the nutrient solution temperature at the same temperature as the air surrounding the aerial portion of the plant. If the nutrient solution is cooler, plant wilting is likely to occur; and if higher than the ambient temperature, physiological root functions, particularly for exposed roots (aeroponics), will be impaired, thus affecting the absorption of water and some of the essential elements.

Root absorption of ions from a nutrient solution requires energy generated by root respiration. For absorption to occur, the roots must be in an aerobic atmosphere; that is, $O_2$ must be present. With some hydroponic growing systems, particularly ebb-and-flow, NFT, and when the applied nutrient solution must flow down an extended column of rooting medium, the lack of $O_2$ during flooding and at the end of a flow of nutrient solution may restrict water and ion absorption. This restriction of absorption can slow plant growth and may even result in plant nutrient element insufficiencies. Therefore, preventing possible anaerobic conditions from developing around roots and/or within the rooting environment is essential.

The lack of $O_2$ in the rooting environment is a major cause of root death. To ensure that a nutrient solution is air or $O_2$ saturated, some recommend the bubbling of air or $O_2$ through the nutrient solution before recirculating.

### 23.7.10 Nutrient Solution Elemental Content Determination and Monitoring

A prepared nutrient solution should be assayed to confirm its elemental concentration before use, as assuming that the elemental concentration of a prepared nutrient solution will always be based on the reagents used and their weight per volume of water is insufficient because errors in the identification of the reagent and amount weighed are not uncommon. Injection pumps used to feed nutrient solution concentrates into a flow of water begin to wear, easily get out of adjustment, and malfunction, thereby affecting the composition of the delivered nutrient solution.

### 23.7.11 Use Factors

The use of a nutrient solution has four parameters:

1. Mineral elemental content and concentration
2. Volume applied at each irrigation
3. Frequency of application
4. Number of plants

These four factors are interdependent and vary considerably depending on the hydroponic system used and the plant requirements described for each of the six hydroponic growing systems.

## 23.8 REAGENTS AND NUTRIENT SOLUTION FORMULATIONS

Reagents for preparing concentrates are typically of higher purity than fertilizers used in field-grown situations. The selected reagents must be water soluble and should be relatively free of substances that are not water soluble or will add unwanted elements. Some of the common reagent sources used as major element sources are magnesium sulfate or Epsom salts, calcium nitrate, monopotassium phosphate, ammonium nitrate, and potassium nitrate. Micronutrient sources for Cu, Fe, Mn, and Zn are their sulfate or chloride salts. A list of reagents for making a nutrient solution is provided in Table 23.5.

A nutrient solution is made by dissolving required element-containing reagents in water, in an amount necessary to achieve a certain elemental concentration. Most nutrient solution formulations are based on what was proposed by Hoagland and Arnon in their 1950 circular (Hoagland and Arnon, 1950) (Table 23.5).

A common means for formulating a nutrient solution is to prepare concentrates of certain groups of elements, known as stock solutions, frequently designated A, B, and C, etc., mixing with water portions of each stock solution to obtain the desired

## TABLE 23.5
## Reagents and Quantity Used to Make a Hoagland and Arnon (1950) Nutrient Solution

| Stock Solution | To Use (ml/L) |
|---|---|
| **Solution No. 1 (without ammonium):** | |
| 1M potassium dihydrogen phosphate (KH$_2$PO$_4$) | 1.0 |
| 1M potassium nitrate (KNO$_3$) | 5.0 |
| 1M calcium nitrate [Ca(NO$_3$)$_2$·4H$_2$O] | 5.0 |
| 1M magnesium sulfate (MgSO$_4$·7H$_2$O) | 2.0 |
| **Solution No. 2 (with ammonium):** | |
| 1M ammonium dihydrogen phosphate (NH$_4$H$_2$PO$_4$) | 1.0 |
| 1M potassium nitrate (KNO$_3$) | 6.0 |
| 1M calcium nitrate [Ca(NO$_3$)$_2$·4H$_2$O] | 4.0 |
| 1M magnesium sulfate (MgSO$_4$·7H$_2$O) | 2.0 |

| Micronutrient Stock Solution | mg/L |
|---|---|
| Boric acid (H$_3$BO$_3$) | 2.86 |
| Manganese chloride (MnCl$_2$·4H$_2$O) | 1.81 |
| Zinc sulfate (ZnSO$_4$·5H$_2$O) | 0.22 |
| Copper sulfate (CuSO$_4$×5H$_2$O) | 0.08 |
| Molybdate acid (H$_2$MoO$_4$·H$_2$O) | 0.02 |
| To use: 1 mL/L nutrient solution | |

| Iron | |
|---|---|
| For Solution No. 1: 0.5% iron ammonium citrate: | To use: 1 mL/L nutrient solution |
| For Solution No. 2: 0.5% iron chelate: | To use 2 mL/L nutrient solution |

elemental concentration in the final plant-delivered nutrient solution. These concentrates can be slowly injected into a flowing water stream simultaneously to create the final nutrient solution.

Commercially prepared nutrient solution concentrates (frequently referred to as stock solutions) are readily available; concentrates when diluted with water will generate a particular formulation for delivery to plant roots, thereby supplying the required essential elements for normal plant growth. Most concentrates do not contain all the essential plant nutrient elements as some elements in concentrated form are not compatible. It may take two or three element concentrates (stock solutions) when diluted and mixed together to obtain a thirteen element-containing complete nutrient solution. One needs to carefully read the label to determine what elements are included and at what concentration exists a concentrate. A common error is to mix concentrates that contain the same element or elements.

Some concentrates whose final formulations are designed for use at a particular stage of growth as the plant advances through its life cycle from the seedling stage to early and mid-maturity, and then to flowering and fruit formation, are of questionable value. The need for a growth-related nutrient solution formulation is questionable when factors such as frequency and volume applied, nutrient solution retention, and elemental accumulation in the rooting medium are the significant influencing factors that will determine elemental availability and plant nutritional status.

A nutrient solution can be made by the grower. Those reagents, and their elemental content, commonly used to make a nutrient solution are given in Table 23.5.

## 23.9 CONCENTRATION RANGES AND RATIOS

All thirteen essential mineral plant nutrient elements in a nutrient solution must be within a particular concentration range as well as in a particular ratio among certain elements in order to ensure sufficiency. Most nutrient solution formulations contain higher concentrations of some elements than required by the plant, P in particular as well as N. In some nutrient solution formulations, the ratios among the major cation elements—$K^+$, $Ca^{2+}$, and $Mg^{2+}$—are frequently out of balance.

Elements in a nutrient solution interact among themselves, exhibiting both antagonistic as well as synergistic characteristics. For example, among the major cations ($K^+$, $Ca^{2+}$, and $Mg^{2+}$), the least competitive is $Mg^{2+}$; therefore, its deficiency is likely to occur with the use of some nutrient solution formulations having high concentrations of $K^+$ and/or $Ca^{2+}$ when growing Mg-sensitive plants such as tomato. The ammonium-N ($NH_4^+$) cation is a strong competitor and can reduce the uptake of both Ca and Mg, resulting in a deficiency in either element for those plant species sensitive to either Ca or Mg or both. For example, a high concentration of $NH_4^+$ in a nutrient solution can result in a high incidence of blossom-end rot in tomato and pepper fruits. There is a synergistic relationship between nitrate nitrogen ($NO_3$-N) and K, the presence of high concentrations of $NO_3$-N enhancing the uptake of K. The presence of a low concentration of $NH_4$-N in a nutrient solution will enhance the uptake of $NO_3$-N.

The elemental concentration ranges for commonly used nutrient solution formulations are given in Table 23.6, and, if the Hoagland/Arnon formulations are used, in Table 23.7.

## 23.10 pH INTERPRETATION-HYDROPONIC NUTRIENT SOLUTION

The optimum pH range for best plant growth is between 5.0 and 6.0, although there are some plant species that tend to favor the pH being closer to pH 6.0. With the pH increasing above 6.0, precipitation and chelation can take place, resulting in essential element insufficiencies. In most hydroponic growing systems with plant roots in continual contact with the nutrient solution, its pH will normally become acidic with time, although depending on the nutrient solution formulation, method of growing, and plant species, the pH may drift upward with time. Therefore, recirculation of the nutrient solution requires periodic pH adjustment in order to maintain constancy at that pH best suited for the hydroponic growing method, nutrient solution formulation, and plant species.

## TABLE 23.6
## Major Element and Micronutrient Ionic Forms and Normal Concentration Ranges Found in Most Nutrient Solutions

| Element | Ionic Form | Concentration Range mg/L (ppm)[a] |
|---|---|---|
| **Major Elements** | | |
| Nitrogen (N) | $NO_3^-$, $NH_4^+$ | 100–200 |
| Phosphorus (P) | $HPO_4^{2-}$, $H_2PO_4^{-}$[b] | 15–30 |
| Potassium (K) | $K^+$ | 100–200 |
| Calcium (Ca) | $Ca^{2+}$ | 200–300 |
| Magnesium (Mg) | $Mg^{2+}$ | 30–80 |
| Sulfur (S) | $SO_4^{2-}$ | 70–150 |
| **Micronutrients** | | |
| Boron (B) | $BO_3^{3-}$[c] | 0.30 |
| Chlorine (Cl) | $Cl^-$ | |
| Copper (Cu) | $Cu^{2+}$ | 0.01–0.10 |
| Iron (Fe) | $Fe^{3+}$, $Fe^{2+}$[d] | 2–12 |
| Manganese (Mn) | $Mn^{2+}$ | 0.5–2.0 |
| Molybdenum (Mo) | $MoO_4^-$ | 0.05 |
| Zinc (Zn) | $Zn^{2+}$ | 0.05–0.50 |

[a] Concentration range based on what is found in the current literature.
[b] Ionic form depends on the pH of the nutrient solution.
[c] The molecule boric acid ($H_3BO_3$) can be absorbed by plant roots.
[d] Ionic form depends on the pH and oxygen level in the nutrient solution.

## 23.11 RECONSTITUTION OF THE NUTRIENT SOLUTION

To maximize use of water and the added essential plant nutrient elements, recirculation of the nutrient solution is practiced; but recirculation requires reconstitution to maintain the pH and elemental composition, filtering to remove suspended materials, and sterilization to kill disease organisms.

## 23.12 ACCUMULATION OF NUTRIENT ELEMENTS AND PRECIPITATES

With the use of a rooting medium (such as gravel, sand, perlite, rockwool, coir, etc.), the medium itself begins to change due to the accumulation of unspent essential plant nutrient elements, either remaining as ions in solution or eventually accumulating as precipitates (mainly a mixture of calcium phosphates and sulfate that, in turn, co-precipitate with other elements, such as Mg and the micronutrients, Cu, Fe, Mn,

**TABLE 23.7**

**Elemental Concentration in the Hoagland and Arnon (1950) Nutrient Solution Formulations**

| Nutrient Element | Hoagland No. 1 | Hoagland No. 2 (ppm) |
|---|---|---|
| Nitrogen ($NO_3$-N) | 242 | 220 |
| Nitrogen ($NH_4$-N) | — | 12.6 |
| Phosphorus (P) | 31 | 24 |
| Potassium (K) | 232 | 230 |
| Calcium (Ca) | 224 | 179 |
| Magnesium (Mg) | 49 | 49 |
| Sulfur (S) | 113 | 113 |
| Boron (B) | 0.45 | 0.45 |
| Copper (Cu) | 0.02 | 0.02 |
| Iron (Fe) | 7.0 | 7.0 |
| Manganese (Mn) | 0.50 | 0.50 |
| Molybdenum (Mo) | 0.010 | 0.010 |
| Zinc (Zn) | 0.48 | 0.48 |

and Zn). These effects can be modified by reducing the element content concentration of the nutrient solution and increasing the volume applied with each irrigation and alternating between the application of the nutrient solution and water. When the retained solution in the rooting medium reaches a certain EC, the rooting media will require water leaching to remove the accumulated ions. However, water leaching will not remove any accumulated precipitates. After one crop use, the rooting medium is normally discarded, an added expense when these methods of hydroponic growing are used.

# 24 Soilless Rooting Growing Media

Organic soilless mixes were developed to replace mineral soils as a potting (rooting) medium. Today, a range of rooting medium mixes are marketed to commercial growers and home gardeners, each mix formulated to suit a particular use, such as for seed germination and seedling production, short-term use for transplant production, and for the long-term growth of plants in containers in indoor environments. Various aspects of this topic are also discussed in Chapter 25, "Organic Farming/Gardening" and Chapter 23, "Hydroponics."

## 24.1 SOILLESS MEDIA INGREDIENTS

Soilless mixes usually contain two or more primary ingredients; both their individual properties and combination effect determine the physical characteristics of the mix. Other materials or substances are added to give a specific physical character to the final mix as well as supply essential plant nutrient elements.

**Major ingredients** used to formulate a soilless mix include the following:

1. *Sphagnum peat moss:* Commonly called peat moss; a naturally occurring substance found in what are called bogs primarily located in northern climatic zones, the result of accumulated, partially decomposed plant (between 151 and 350 species of mosses) materials whose characteristics are determined by the source material and decomposition conditions existent during its formation. The harvested material may be crushed and screened to obtain a particular particle size. The material has good physical stability, relatively high water-holding capacity, a cation exchange capacity, and may contain some essential plant elements (see Table 23.2)
2. *Pinebark (composted, milled):* A by-product from the lumbering of pine tress, the removed bark is crushed and composted, composting time and conditions determining the characteristics of the bark. Composting is necessary to lower the tannin (tannins are toxic to plants) content. Depending on the type of bark and particle size, water-holding capacity varies; intercellular space will adsorb and hold water; pinebark may be slow to wet, is highly acidic, and may contain toxic levels of Mn. Particle size will determine its water-holding and aeration characteristics (see Table 23.2).
3. *Core:* A by-product of coconut production that is coming into wider use as a substitute for either sphagnum peat moss or pinebark. There are a variety of core products, including blocks and slabs that have specific usage for the germination of long-term growth of plants (see Table 23.2)

4. *Vermiculite:* A naturally occurring mineral having a fairly high water-holding capacity, high cation exchange capacity, and can be a source of K, and/or Mg and Fe, three essential plant nutrient elements.
5. *Perlite:* A naturally amorphous volcanic glass that is formed by heating to 850 to 900°C; the expended material is a brilliant white and lightweight (see Table 23.1).
6. *Sawdust:* A by-product in the production of lumber, the properties of which are determined by tree species. Fresh sawdust is not suitable for use, requiring composting to reduce the content of tannins that can be toxic to plants (see Table 23.2).
7. *Compost:* An end product from the composting of substances from a wide range of sources, having properties that are determined by the source materials and degree of decomposition.
8. *Animal manure:* Source and degree of composting determines its physical and chemical properties, and it may contain significant quantities of essential elements, such as N, P, and K, as well as some micronutrients, and therefore may be added to a soilless mix to supply these essential plant elements (see page 168).
9. *Worm castings:* End product from the decomposition of an organic material by worms. The type of organic material will determine, to some degree, the physical and chemical properties of the final product (see page 169).

**Supplemental substances** include the following:

1. *Liming material:* Added to:
   - Establish and maintain the desired pH of the soilless mix.
   - Counter the acidity of the major ingredients.
   - Provide a source of Ca as well as Mg for plant use.
   - The source and form of the liming material (see Chapter 18, "Lime and Liming Materials") added is determined by the desired effect.
2. *Calcium sulfate (gypsum):* Added to a mix to provide a source of Ca that would not be adequately supplied by an added liming material, for mixes without a liming material added, or a source of Ca that will not influence its pH.
3. *Inorganic fertilizer:* Added to provide a source of essential elements, either in a liquid or solid chemical form (see Chapter 19, "Inorganic Fertilizers and Their Properties"), or organic form (see Chapter 20, "Organic Fertilizers and Their Properties"), the selection being made based on use for the soilless mix and requirement for the essential plant elements during use.
4. *Wetting agent:* added to a soilless organic mix to make the mix easy to wet.
5. *Sand:* Can be a major ingredient to give weight to the soilless mix and/or to establish a particular drainage characteristic.

## 24.2 SOILLESS MEDIA FORMULATIONS

The first major soilless mix formulations were the Cornell Soilless Peat-Lite Mixes, a product of research and development by researchers at Cornell University. They developed three organic soilless mixes based on the combination of sphagnum peat with the other major ingredient being either vermiculite or perlite:

| Ingredients | Amount |
|---|---|
| **Peat-lite Mix A** | |
| Sphagnum peat moss | 11 bu |
| Horticultural vermiculite No.2 | 11 bu |
| Limestone, ground | 5 lbs |
| Superphosphate, 0-20-0 | 1 lb |
| 5-10-5 fertilizer | 2–12 lbs |
| **Peat-lite Mix B** | |
| Sphagnum peat moss | 11 bu |
| Horticultural perlite | 11 bu |
| Limestone, ground | 5 lbs |
| Superphosphate, 0-20-0 | 1 lb |
| 5-10-5 fertilizer | 2–12 lbs |
| **Peat-lite Mix C (for germinating seed)** | |
| Sphagnum peat moss | 1 bu |
| Horticultural perlite | 1 bu |
| Limestone, ground | 7.5 oz |
| Superphosphate, 0-20-0 | 1.5 oz |
| Ammonium nitrate | 1.0 oz |

*Note:* For Cornell Peat-lite Mixes A and B, additional ingredients may be added to supply the essential elements, B and Fe: 10 g of borax and 25 g chelated Fe, if needed.

*Source:* Sheldrake, R., Jr. and J.W. Boodley. 1965. *Commercial Production of Vegetable and Flower Plants.* Cornell Extension Bulletin 1065. Cornell University, Ithaca, NY.

For the short-term growth of plants, additional mixes were developed by researchers in California and Canada, both mixes using sphagnum peat moss as the major ingredient:

| Ingredients | Amount |
|---|---|
| **University of California #E** | |
| Sphagnum peat moss | 22 bu |
| Limestone, ground | 7.5 lbs |

| Superphosphate, 0-20-0 | 1.0 lb |
| Potassium nitrate | 0.3 lbs |

### Canadian Seedling Mix

| Sphagnum peat moss | 12 bu |
| Horticultural vermiculite | 10 bu |
| Limestone, ground | 4 lbs |
| Superphosphate, 0-20-0 | 1 lb |
| 10-10-10 fertilizer | 2 lbs |
| Potassium nitrate | 0.5 lbs |
| Borax | 1.0 g |

All these mixes were primarily used for either the germination of plants from seed, seedling production, or the short-term growth of ornamental plants. For long-term growth, these mixes will require nutrient element supplementation, the kind, amount, and frequency of application determined by the growth requirement of the plants.

For the long-term production of greenhouse tomatoes, researchers at Rutgers University based their soilless organic mix essentially on the Cornell Peat-Lite Mix A:

| Ingredients | Amount |
| --- | --- |
| Sphagnum peat moss | 9 bu |
| Horticultural vermiculite | 9 bu |
| Limestone, ground | 8 lbs |
| Superphosphate, 0-20-0 | 2 lbs |
| Calcium nitrate | 1 lb |
| Borax | 10 g |
| Chelated Fe | 35 g |

Researchers at the University of Georgia used milled composted pinebark as the major ingredient—rather than sphagnum peat moss—for growing greenhouse tomatoes:

| Ingredients | Amount |
| --- | --- |
| Milled composted pinebark | 9 bu |
| Limestone, ground | 1 lb |
| Fertilizer, 10-10-10 | 1 lb |

Both mixes required periodic plant nutrient element supplementation based on either plant analysis or on plant growth and appearance.

All these mixes illustrate the variety of formulations that have been used, with some still in wide use today for the growing of plants under a range of growing conditions. Horticultural-grade vermiculite is not a common ingredient in most soilless mixes today; the major ingredients are sphagnum peat moss and perlite with some soilless organic mixes, including composted milled pinebark as a supplement to sphagnum peat moss.

## 24.3 PHYSICAL PROPERTIES

The physical properties of a mix, the texture (particule size and distribution), volume-weight, and water-holding capacity will be determined by the physical properties of the major ingredients and their relative proportions in the mix.

## 24.4 PHYSIOCHEMICAL PROPERTIES

The pH, cation exchange capacity, and level of elemental availability will be determined by the characteristics of the major ingredients as well as the characteristics and quantity of the supplemental substances added, such as liming material, chemical inorganic fertilizer, organic fertilizer, etc.

## 24.5 CONTROL OF pH

Keeping the water pH of an organic mix less than 5.5 is essential to prevent potential micronutrient deficiencies from occurring, mainly Cu, Fe, Mn, and Zn, because the availability of these elements decreases with increasing pH. If hard water is used for irrigation, less limestone (dolomitic or otherwise) should be initially added to the mix because there may be sufficient Ca and possibly Mg in the water to satisfy the crop requirement.

## 24.6 USE FORMULATIONS

1. *Germination and seedling production:* A fine particle mix that can hold water and provide close contact between the seed and seedling roots is required.
2. *Short-term growing:* A particle mix with moderate water-holding capacity and containing sufficient essential elements to meet the demand of plants prior to their transfer into another rooting environment.
3. *Long-term growing:* A coarse mix is best with moderate water-holding capacity with sufficient air space to keep roots actively respiring.

## 24.7 BAG CULTURE SYSTEMS

Prior to currently used hydroponic growing systems, soil was the rooting medium in tomato greenhouses. Because management of the physical and chemical properties of the soil was increasingly becoming a major problem, the next innovation tried was bag culture using some form of customized mix. The bags were placed on the greenhouse floor and a tomato plant was placed in an opening in each mix bag. Various organic substances, such as peat moss, vermiculite, perlite, humus, and wood products, were in the mix to give it certain physical and chemical properties. After use, the bags of mix could be discarded, thus limiting the possible occurrence of disease problems. The use of soilless bag culture for the production of greenhouse tomatoes had a short history because hydroponic growing systems were being introduced and put into use.

Two systems were used, one in which all the needed nutrient elements were in the mix or, as the season progressed, needed nutrient elements would be added through the applied irrigation water as needed. The ingredients in a typical soilless bag mix for tomato would be as follows:

### Ingredients for a Complete and Base Mixture of Peat Moss

| Ingredient | Complete | Base |
|---|---|---|
| Peat moss | 0.5 m³ | 0.5 m³ |
| Horticultural vermiculite | 0 5 m³ | 0.5 m³ |
| Ground limestone (dolomite) | 7.5 kg | — |
| Limestone (pulverized FF) | — | 5.9 |
| Gypsum (calcium sulfate) | 3.0 kg | — |
| Calcium nitrate | 0.9 kg | — |
| Superphosphate, 20% P | 1.5 kg | 1.2 kg |
| Epsom salts (magnesium sulfate) | 0.3 kg | 0.3 kg |
| Osmocote 18-6-12 (9 months) | 5–6 kg | — |
| Chelated iron (NaFe 138 or 330), 10% | 30 g | 35 g |
| Fritted trace elements (FTE 503) | 225 g | 110 g of FTE 302) |
| Borax (sodium borate) | 35 g | |

Another organic mix formulation for growing tomatoes was an equal ratio for peat moss and horticultural vermiculite or an equal mix of peat moss, horticultural vermiculite, and perlite placed in a 42-L plastic bag that measures 35 ´ 105 cm when flat. Dolomitic limestone and various fertilizers (i.e., superphosphate, potassium nitrate, calcium nitrate, magnesium sulfate, and micronutrients) were added to these mixes to supply the needed essential nutrient elements.

A formula for a tomato bag mix of peat nodules (sedge or humified sphagnum) in a 20-L bag is as follows:

| Ingredient | kg/m³ | lbs (oz)/yd³ |
|---|---|---|
| Superphosphate (0-20-0) | 1.75 | 3 lbs |
| Potassium nitrate | 0.8 | 7 lbs 8 oz |
| Potassium sulfate | 0.44 | 12 oz |
| Ground limestone | 4.2 | 7 lbs |
| Dolomitic limestone | 3.0 | 5 lbs |
| Frit 253A | 0.4 | 10 oz |

*Note:* Additional slow-release nitrogen (N) at 0.44 kg/m³ urea-formaldehyde (167 mg N/L) is sometimes included. If slow-release phosphorus fertilizer is required, magnesium ammonium phosphate (MagAmp or Enmag) at 1.5 kg/m³ is added.

Another bag organic mix suitable for tomato culture consists of the following ingredients:

| Ingredient | Amount |
|---|---|
| Sphagnum peat moss | 9 bu |
| Vermiculite | 9 bu |
| Perlite | 9 bu |
| Dolomitic limestone | 8 lbs |
| Superphosphate fertilizer (0-20-0) | 2 lbs |
| Calcium nitrate | 1 lbs |
| Borax | 10 g |
| Chelated iron | 35 g |

Greenhouse tomatoes have been successfully grown in pure milled pinebark in a growbox system that was suitable for home garden use. A 7-inch depth of bark is placed in a watertight box and an inch of water is maintained in the bottom of the box. All the nutrient elements required by the tomato plant are initially mixed into the pinebark using dolomitic limestone and 10-10-10 fertilizer.

| Ingredient | Amount |
|---|---|
| Milled pinebark | 9 bu |
| Dolomitic limestone | 1 lb |
| Fertilizer, 10-10-10 | 1 lb |

With extended use, additional fertilizer must be added to maintain a tomato plant in good nutritional status. A micronutrient mix was not needed as there was usually a sufficiency of these elements as contaminants in the various ingredients.

A major requirement for bag culture is control of watering to ensure sufficiency but not excess. For a peat moss bag system, it has been found that restricting water to 80% of the estimated requirement resulted in a 4% loss in yield (smaller fruit) but improved fruit flavor. It has been found that high water levels resulted in Mn deficiency. In addition, bag size can influence fruit yield and the incidence of blossom-end-rot (BER), with 7-L bags producing lower yields and high BER incidence, while 14-, 21-, and 35-L bags produced about the same yield with a lower incidence of BER.

Although some growers are still using an organic mix, either in bags or pots, there is little interest today in developing modifications of these mixes to improve their performance under greenhouse growing conditions for the production of tomatoes.

Therefore, there is little future for this method of greenhouse tomato production. The major limitation of bag or pot soilless mix culture systems is the inability to carefully control the moisture supply and nutrient element level that is required for the production of high fruit yields.

## 24.8  FERTILITY DETERMINATION PROCEDURE FOR AN ORGANIC SOILLESS MIX

The standard testing procedures for the determination of the fertility status of a mineral soil do not lend themselves to the assay of an organic soilless mix. The procedure used is by water-equilibrium extraction (Jones, 2001), a procedure that is commonly used in most soil testing laboratories. Pure water is stirred into an aliquot of organic soilless medium until there remains a film of free water on the surface of the medium. The slurry is vigorously stirred for several minutes and then allowed to stand for 1 hour. If there is no film of free water on the surface, additional water is added until free water appears on the surface. If there is some depth of water on the surface, add more organic soilless medium and stir, adding medium until there is just a film of free water on the surface. Let the slurry stand for an additional hour. Stir and with continuing stirring, place the electrodes attached to a calibrated pH meter into the slurry and determine the pH. The optimum pH range for most organic soilless mixes is between 5.0 and 5.6.

Removing the water from the slurry by filtering through quality filter paper to gives a clear filtrate. Collect the filtrate for elemental analysis. A soluble salt measurement is made in the filtrate.

The soluble salt level and element content in the filtrate are used to define the fertility status of the organic soilless mix:

## Level of Acceptance

| Determination | Low | Acceptable | Optimum | High | Very High |
|---|---|---|---|---|---|
| Soluble salts, dS/m | 0–0.75 | 0.76–2.0 | 2.1–3.5 | 3.6–5.0 | 5.1+ |
| Nitrate-N, mg/L | 0–39 | 40–99 | 100–199 | 200–299 | 300+ |
| Phosphorus, mg/L | 0–2 | 3–5 | 6–10 | 11–18 | 19+ |
| Potassium, mg/L | 0–59 | 60–149 | 150–249 | 250–349 | 350+ |
| Calcium, mg/L | 0–79 | 80–199 | 200+ | | |
| Magnesium, mg/L | 0–20 | 30–69 | 70+ | | |

# Section VII

---

## Miscellaneous

# 25 Organic Farming/ Gardening

The concept of organic crop production has its roots in the Humus Theory, one of several theories developed in the 1880s by investigators as a possible explanation as to "how plants grow." This concept stated that *Mother Earth* was the "food" source for plants, an explanation that had considerable support among scientists at that time. In early field experiments, it was observed that plants grown on soils treated with organic manures grew better than those receiving various inorganic substances. In addition, plants growing in mineral soils that were high in organic matter content grew more vigorously than those in mineral soils of low organic content. As scientists began to identify which mineral elements were essential in order for plants to grow normally, some took the next step by suggesting that the form (i.e., organic or inorganic origin) of these newly identified essential elements taken into the plant by root absorption is a factor in determining plant health. The next step that has occurred more recently is the belief that the plant whose elemental composition is organically derived will produce fruit and grain healthier than these same products taken from plants whose essential elementals were root absorbed from inorganic sources, such as chemically made fertilizers. Therefore, those consuming these healthier foods will then receive the same benefit in terms of one's own health and well-being. There has not been sufficient research to verify that such does occur, suggesting that ions derived from an organic source have properties different from those that were derived from an inorganic source. More recently, the concept of organic farming/gardening has taken on a wider scope to also include all man-made chemicals used to control weeds, insects, and diseases.

## 25.1 CHEMICALIZATION OF CROP PRODUCTION

Beginning in the late 1940s, the chemicalization of crop production practices began to significantly increase, initially with the use of pest control chemicals, followed by chemically formulated fertilizers and herbicides. There arose concern on what effects chemicals designed to protect plants from insects and diseases would have on the environment. In addition, most of these chemicals required the applicator to use protective clothing as well as specialized storage and handling procedures to ensure user safety.

The environmental issue was brought to the public's attention with the publication of Rachel Carson's 1962 book, *Silent Spring*. The premise of her book was that with increasing use of crop protection chemicals, mainly insecticides, adverse side effects were beginning to be observed in the surrounding environment, resulting in biological imbalances among species of insects, birds, and animals. In addition, there arose concern for the health safety of food products in the marketplace if they carried residues of applied crop-protection chemicals as many of these chemicals had long half-lives, with their presence and toxicity remaining in the environment for long periods of time.

The counter argument was that with the increasing concentration of one species of plants in a relatively small area, there would be the natural occurrence of an accompanying insect and disease prevalence, and therefore a "created imbalance" that could be brought back into balance by the use of pest control chemicals impacting only the area where they were being applied. In addition, without chemical control, crop yield and quality would be adversely affected, losses that would reduce the supply and increase the cost of primarily fruits and vegetables to consumers, important dietary food items that contain essential vitamins and minerals.

As the movement by a concerned public to either significantly reduce or even remove from use crop-protection chemicals gained momentum, the name given to this movement was first identified as organic gardening and then organic farming. By the mid-1990s, there arose considerable consumer demand for food products that had been grown without using man-made chemicals. Initially, organically grown fruits and vegetables were of low quality, and in many instances, lacked good eye appeal. However, as growers learned how to grow using organic substances and organic-based procedures, both product quality and availability increased, although cost still remains higher than the same item grown using chemically made chemicals in their production. To regulate what can be identified as organically grown, both state and federal regulations began to be issued.

There has been insufficient research or experience to substantiate the concept that consuming organically grown food products will affect one's physical health for the better, or that consumption of nonorganically produced products will affect one's health.

## 25.2 "ORGANICALLY GROWN" DEFINED

Over the past decade, criteria have been developed that define organic crop production requirements. Today there is a national system of designated requirements in order for food products to be marketed as "organically-produced;" there are state regulations as well. State programs specify the required procedures for the production of designated organically grown food products, having the responsibility for handling the application and inspection procedures as well as monitoring for compliance.

In today's markets, organically grown food products are readily available, of reasonable quality (have good eye appeal), but at a cost higher than the same food product grown by other than organic means. For some consumers, cost and appearance factors are of no concern, the influencing factors being safety and nutritional value assumed to be associated with the designation as being "organically grown."

## 25.3 SUITABLE INORGANIC FERTILIZERS

Today, growing food plants organically requires selecting those sites that have not been recently treated with man-made chemicals, and then producing a crop without the use of any man-made chemicals, including most commercial fertilizers.

Some of the commonly used fertilizers are naturally occurring substances, such as potassium chloride (KCl), commonly known as muriate of potash, or just potash, which occurs in natural deposits; and potassium sulfate ($K_2SO_4$), also mined from natural deposits or obtained from salt brines. Most liming materials, such as calcitic and dolomitic limestone, are naturally occurring substances. Rock phosphate, a naturally occurring mineral, was in the past the primary fertilizer source for P (see page 165). Today, most P-containing fertilizers are chemically made products (see page 159). Below is a partial listing of those substances identified as "organically acceptable":

| Substance | Elemental Content or Composition |
|---|---|
| Calcite, calcitic limestone | 95% to 100% calcium carbonate |
| Colloidal phosphate or soft | 0% N, 18% to 20% P, 27% Ca, 1.7% iron phosphate silicas, 14 other trace elements |
| Dolomite, dolomitic limestone | 51% calcium carbonate, 40% magnesium carbonate |
| Granite dust, granite meal | 0% N, 0% P, 3% to 5% K, 67% silica, 19 trace crushed granite minerals |
| Green sand, glauonite | 0% N, 10% P, 5% to 7% K, 50% silica, 18% to 20% iron oxide, 22 trace minerals |
| Gypsum (calcium sulfate) | 23% to 57% C, 17.7% S |
| Langbeinite | 0% N, 0% P, 22% K, 22% S, 11% Mg |
| Rock phosphate | 0% N, 22% P, 0% K, 30% Ca, 2.8% Fe, 10% silica, 10 other trace minerals |
| Sulfur (elemental) | 100% S |

The reaction of these and similar substances in soil depends on their physical and chemical properties as well as the physiochemical properties of the soil

| Fertilizer Material | Estimated N-P-K | Rate of Nutrient Release |
|---|---|---|
| Granite meal | 0.0–0.0–4.5 | Very slow |
| Green sand | 0.0–1.5–5.0 | Very slow |
| Colloidal phosphate | 0.0–16.0–0.0 | Slow–Medium |
| Rock phosphate | 0.0–18.0–0.0 | Very slow–Slow |
| Soybean meal | 6.5–1.5–2.4 | Slow–Medium |
| Wood ash | 0.0–1.5–5.0 | Fast |

## 25.4   SUITABLE ORGANIC FERTILIZERS

Finding suitable organic fertilizer sources for the major essential plant elements, particularly N and P, is a challenge. Most N- and P-containing fertilizers are chemically made (see page 159). Before fixed-N fertilizers became available (after World War II using the Haber process that was developed for production of munitions), animal manures and green-manure legume crops were the primary sources of N. Rotational grain crop production systems would include a legume crop that would be the source of N required by the grain crop. Today, continuous grain production is possible with the use of chemically generated N sources (see pages 159).

## 25.5   ORGANIC SOIL FERTILITY MANAGEMENT

An elementally infertile soil can be quickly restored to optimum fertility status by applying soil test-based applications of lime and chemical fertilizers. Applying organic materials, such as manures, composts, organic waste products, etc., of unknown and widely varying elemental contents, frequently lacking elements in greatest need, or containing elements not needed, can create elemental insufficiencies, thereby adding to the soil's infertile status. This is particularly true for the home gardener or same acreage farmer, who can easily overdose his garden or field soil by not taking into account the plot size and rate of application of an organic material of unknown elemental content, thus creating an elemental imbalance that can be difficult to correct.

## 25.6   SOIL PHYSICAL PROPERTIES

Applying organic materials, such as manures, composts, organic waste products, etc., can improve the soil's physical properties, increasing water-holding capacity, improving soil structure, tilth, and degree of aeration, properties that can add to the soil's fertility status. Overdosing a mineral soil with these same organic materials, however, will significantly change the physical and chemical properties of the soil that then requires a different set of management practices to sustain its fertility status. High (>10%) organic matter content soils are much more difficult to manage than mineral soils of lower organic matter content. High organic content soils tend to stay wet after rain or irrigation, are cooler, and therefore plant growth will be slower; and the microorganism populations will be high, becoming competitors for essential plant elements and possibly a source for organisms that can become pathogenic to plant roots.

## 25.7   FOOD SAFETY AND QUALITY ISSUES

Another danger is that failing to either compost or sterilize animal manures, or other similar organic by-products, can increase the risk for the occurrence of *salmonella* and *Escherichia coli* bacterial illnesses, particularly when applied to soils used to grow vegetable crops. It should be remembered that composting an organic material results in the loss of organic components, making the composted material considerably higher in elemental content than what was in the original material.

The suggestion that an organically produced plant product has higher nutrient value than that same product inorganically produced cannot be easily verified because there are many other factors that can influence nutrient value, such as natural soil physical and chemical properties, air and soil temperatures, soil moisture, factors associated with plant characteristics, and applied cultural practices. Plant genetics is now becoming an issue because genetically modified crop plants are being more widely grown as they were bred and selected in order to obtain resistance to pests and are adapted to environmental stresses.

# 26 Weather and Climatic Conditions

Current and past weather conditions can have a significant effect on plant growth, product yield, and product quality. Air temperature sequences, rainfall amounts and distribution, cloud cover, minutes of sunshine, and wind are records that must be examined when dealing with long-term or developing crop anomalies. Short and long periods of weather extremes will affect plant performance. Consistency of weather conditions can be as stressful on a plant as the cycling of weather conditions. Plant appearance at any point in time may be the result of earlier weather conditions, rather than current conditions. Previous-year weather conditions can have a carryover effect on plant growth in the current year, particularly rainfall. The fertility of a soil measured by a soil test will be affected by the previous year's rainfall. A wet spring versus a dry spring will affect a soil test result determination equally.

## 26.1 DEFINITIONS

The terms "weather" and "climate" are related, but they refer to two different subjects; weather denotes those conditions that are influenced by air mass movements, and climate refers to what occurs with latitude and the position of the Earth in relation to the sun. Climate has a fixed aspect reoccurring with time, while weather conditions will change from day to day. There exists unique, repeatable seasonal air temperatures, rainfall patterns and cloud cover that, when known, can be used for site selection and cropping patterns. Frequent occurrences of hail, high winds, long drought periods, and air temperature extremes are weather events not suitable for the production of high-value cash crops, and therefore such sites experiencing these conditions should be avoided.

Plant performance at any point in time may be the result of earlier weather conditions, rather than current conditions. Air temperature sequences, rainfall amounts and distribution, cloud cover, minutes of sunshine, and wind are records that must be examined when dealing with long-term or developing crop anomalies. Current and past weather conditions can have a significant effect on plant growth, product yield, and product quality. Short and long periods of weather extremes will affect plant performance. Consistency of weather conditions can be as stressful on a plant as the cycling of weather conditions. Previous-year weather conditions can have a "carry-over" effect on plant growth in the current year, particularly rainfall.

## 26.2    CLIMATIC FACTORS

The five climatic factors requiring consideration are air temperature, rainfall, wind, radiation, and atmospheric carbon dioxide ($CO_2$) content.

### 26.2.1    AIR TEMPERATURE

Most plants grow well in air temperatures that range between 55°F and 90°F, although some plants prefer daytime temperatures more toward the lower portion of this range, while others prefer daytime temperatures at the higher (warmer) end. Air temperature extremes will have a significant effect at all stages of plant growth. Cool daytime air temperatures during initial plant growth significantly slow growth. Nighttime air temperatures affect plant performance as well, with cool nighttime temperatures favoring sustained growth and high fruit quality. Cool nighttime temperatures can result in dew deposition that will be beneficial to a plant, particularly during periods of daytime moisture stress. High nighttime air temperatures during fruit development and maturity will reduce fruit set and decrease fruit quality. Air temperature combined with wet conditions will favor the occurrence of disease incidence, diseases that develop under both warm and cool conditions.

Cool soil temperatures impede seed germination and slow plant growth, even though the air temperature is favorable for good plant growth. The time of planting for many crops is determined by the temperature of the surface soil. Surface soil conditions, such as color, moisture content, plant cover, and roughness are factors that will affect surface soil temperature.

### 26.2.2    RAINFALL

Frequent short periods of rainfall are more favorable to good plant growth, product yield, and product quality than heavy rainfall with long periods between rainfall events. Heavy rainfall can result in the loss of surface soil (erosion), drowning of young plants, and leaching and erosional loss of applied fertilizer, particularly N. An extended period of light rain is most effective for the replenishment of moisture in the rooting soil profile. Rainfall during the period of rapid plant growth may result in plant wilting due to either reduced temperature of the surface soil and/or temporary flooding of the root zone resulting in an anaerobic condition in the root zone.

The rate of movement of rainwater, as well as irrigation water, into and through the soil profile depends on:

- Soil texture
- Soil structure
- Presence and depth to a hardpan
- Physical conditions of the subsoil
- Depth of the surface soil
- Natural internal soil drainage conditions
- Installation of a tile drainage system, condition and functionality
- Depth to the water table

The fertility of a soil measured by a soil test may be affected by the previous year's rainfall. A wet spring versus a dry spring will also affect a soil test result determination. The following soil conditions will occur after

- A long period of reduced rainfall:
  - Accumulation of N as nitrate ($NO_3^-$) anion and B in the rooting zone
  - Increased incidence of Mg deficiency in sensitive crops

- A long period of heavy rainfall:
  - Lower B soil test result
  - Low nitrate ($NO_3^-$) anion soil profile content
  - Increased rate of soil acidity

Drought or wet conditions can trigger the occurrence of an essential plant nutrient element insufficiency due to interference with root absorption and/or upward movement in the transpiration stream within the plant.

## 26.2.3 WIND

Surrounding stagnant air is detrimental to plant growth. Gentle wind around plants keeps plant leaf surfaces cool and dry by sweeping stagnant air off leaf surfaces. Wind passing through a dense plant canopy will reduce disease incidence and maintain $CO_2$ air concentrations at the existing ambient level. For row crops, row orientation can either enhance or block air movement within the plant canopy.

High wind, particularly sustained, can result in physical damage to plants. Hot dry wind increases the rate of water loss from the soil by evaporation and from plant surfaces by transpiration. Temporary plant wilting may occur during periods of hot dry wind during periods of rapid growth even though soil moisture conditions are adequate.

Wind that is natural or mechanically produced during periods at or near freezing air temperatures can reduce frost damage when plants are in their initial stages of growth or during periods of flowering and fruit set.

## 26.2.4 SOLAR RADIATION INTENSITY AND DURATION

Solar radiation intensity and duration influence plant growth and development, whether that from sunlight or generated light. Intermittent periods of either high or low sunlight intensities have a lesser effect on plants than do long periods of either extreme. The occurrence of either of these conditions at critical periods of early plant growth, and/or during flowering, fruit formation, and development are more critical than during the overall vegetative growth period. A plant condition being observed at any point in time related to a light factor is what occurred several weeks previously—thus the need to examine weather records several weeks earlier.

Minutes of sunshine had been the commonly recorded light measurement made at most weather stations. Unfortunately, this determination is no longer being made at many weather stations. It is the deviations from the norm that should be examined when light factor issues are thought to be influencing plant performance. Plant exposure to light: is full sunlight or varying sunlight during the day the ideal? For some plants, the ideal sunlight condition would be full sun during the morning hours, cloudy conditions just before and after high noon, and then full sun in the afternoon.

Sunlight intensity is correlated with air temperature so that the combined effects of high sunlight intensity and high air temperature are the factors that impact the plant with greater force than either alone. Therefore under high light intensity, cool air temperatures may moderate the impact that would have occurred if both high light and high air temperatures occurred simultaneously. Plant effects of high light intensity may not appear during the period of high light intensity, but may appear later, affecting fruit set and quality for a fruiting plant, such as tomato, and may also have an effect on grain yield for crops such as corn.

The other factor influencing plant growth is light duration. Plants do not grow well in continuous light, but an extended period of light can be of significant benefit. During the summer months, plant growth is benefited at the higher latitudes due to longer periods of daylight. Many plants respond to changing hours of daylight, best growth occurring during periods of increasing hours of daylight, and slower growth and lower yields as the daylight hours decline. Those flowering plants that have considerable commercial value are affected by changing light periods because flowering is stimulated by either short or extended light periods.

Solar radiation, its duration, intensity, and wavelength composition are the factors that drive the photosynthesis process (see page 15). That wavelength portion of light radiation (as sunlight or produced) that drives photosynthesis is known as photosynthetically active radiation (PAR).

Water vapor and suspended particles act as a filter, changing both sunlight intensity as well as wavelength distribution; the extent of filtering will affect plant growth.

## 26.2.5   Carbon Dioxide ($CO_2$)

Carbon dioxide ($CO_2$) provides the carbon (C) needed for photosynthesis (see page 16). The normal $CO_2$ content of air ranges between 320 and 340 ppm (parts per million). For some types of plants, the $CO_2$ content of the air surrounding leaves can significantly affect the rate of photosynthesis at both the low and high end of the normal concentration range. At elevated $CO_2$ concentrations, what are known as "C3" plants (see page 235) will have a higher rate of photosynthesis, although even "C4" plants will also be benefited, but to a lesser extent.

The issue of "global warming" centers on the emission of $CO_2$ from the burning of carbon-containing fuels. From a global perspective, this increase has both a potential detrimental as well as beneficial aspect, the beneficial aspect meaning more C as $CO_2$ for utilization by green plant, increasing the rate of photosynthesis, and thereby increasing plant growth.

## 26.3   WEATHER AS A DIAGNOSTIC FACTOR

Either current or prior weather conditions can be the primary cause for the current condition of a crop plant, or their effects will be manifested over the entire season of growth. Previous weather conditions can set in motion growth characteristics that will impact plants over the remaining season. For example, the moisture and temperature conditions during the early growth period of the corn plant will determine ear size and kernel numbers. Plants damaged by hail during early or mid-growth will look "normal" weeks after being damaged, but the reproductive processes will be significantly impaired, resulting in lower than expected yield and quality. Wet and cool weather conditions during early growth of most fruiting crops, such as tomato, will significantly reduce fruit set and fruit quality.

The combination of poor soil fertility conditions and weather stress can result in crop failure, while with "ideal" weather conditions, despite poor soil fertility status, will result in good plant growth and expected yields obtained. In Ohio during years with average rainfall, continuous grown corn without applied N fertilizer gave grain yields equal to the state average, but was less than 25% of the state average during years of below average rainfall. During several years of a continuous drought condition, increased occurrence of Mg deficiency was observed in many different crop plants, which disappeared when the seasonal drought ended. An essential plant nutrient element insufficiency or less than optimum soil pH, will have less effect on a plant when weather conditions are "ideal" for good plant growth than when plants are under a weather stress condition. Trying to correct a plant condition that manifests itself as a suspected essential plant nutrient element insufficiency by means of fertility correction during stressful weather conditions may be ineffective.

# 27 Best Management Practices (BMPs)

Best management practices, better known by the acronym BMPs, were derived as a means for effective, practical, structural or nonstructural procedures to prevent the movement of sediment, nutrients, pesticides, and other pollutants from the land to surface or ground water, thereby protecting water quality. From this beginning, the concept of BMPs has spread to all procedures for the management of soils and crops. Therefore, one could describe BMPs for the growing of corn or vegetable crops on soils of varying physiochemical properties and under a range of climate/ weather conditions.

## 27.1 ORIGIN

The BMP concept was introduced with the passage of The Federal Water Pollution Control Act Amendments of 1972, Public Law 92-500, a law that specified practices designed to control nonpoint water pollution by

- Reducing the amount of runoff
- Reducing the amount of pollution carried by the runoff
- Prohibiting soil-disturbing activities in fragile or severely erosion-prone areas

## 27.2 BEST MANAGEMENT PRACTICE APPLICATION BROADENED

From its environmental beginning, the concept of BMPs has spread to other aspects of the management of soils and crops, as outlined in the Potash & Phosphate Institute booklet (Anonymous, 1991). Today, BMPs focus on management inputs to provide for

- High input efficiency
- Maximum economic yield

while holding to those environmental practices needed to sustain good environmental stewardship by means of soil and water conservation. In addition, BMPs make provision for

- Improving food safety and quality
- Promoting high quality plant production

as well as being sufficiently flexible to accommodate for specific soils, and other growing media and growing systems.

229

One could outline BMPs for the growing of corn or vegetable crops on soils with varying physiochemical properties and ranges of climatic/weather conditions. By incorporating the BMP concept into the farming operation, the farmer follows only those procedures that fit and conform to a designed BMP program based on practices that are continually improved, gained from practical field experience and ongoing research.

## 27.3    BEST PRACTICE

Another word term spun off from best management practice concepts is "best practice," defined as "a technique, method, process, activity, incentive, or reward that is believed to be more effective at delivering a particular outcome than any other technique, method, process, etc. when applied to a particular condition or circumstance (www.Wikipedia.com). What would be the "best practice" for one set of conditions will not always apply to any or all others. This author accompanied a group of farmers from southern Georgia as they toured the vegetable growing areas in California. Although a significant learning experience, many of the vegetable crop growing procedures being used in those California valleys would not be applicable to the soil–climatic conditions existing in southern Georgia.

## 27.4    IMPORTANT PROTOCOL CONSIDERATIONS

It is important to become familiar with those BMP protocols that fit the soil–plant–climatic conditions for the cropping system currently employed. From experience gained with each cropping cycle, a set of procedures, both successful as well as those proven to be unsuccessful, becomes the basis for formulating a set of protocols that becomes the BMP procedures. Monitoring is essential as an input for developing BMP procedures, with the goal of advancing both yield and quality. The most common error is the failure to keep good records, and then not evaluating the final end product in terms of input effects. It is equally important to recognize that there may exist a "glass ceiling" that sets a limit on yield—and possibly quality. Attempting to circumvent such limits may prove costly. Consulting with other farmers, soil–crop–fertility specialists, or similar individuals or groups can bring another perspective that may uncover issues that should be addressed to bring crop performance to the next higher level. Keeping pace with scientific advancements that have application to the existing soil–crop system is an equally important requirement. Just changing one protocol can have a significant effect on crop performance. Matching crop characteristics with soil fertility and other management inputs is important so as to utilize the full potential of each change made.

The challenge today is to be able to distinguish which BMP protocols have universal application from those that are correlated to a particular set of crop–soil–climatic growing conditions. Basic principles have universal application, with modification of a practice being associated with a particular set of correlated conditions as well as the management abilities of the farmer/grower. BMPs should be continually updated, gained through practical field experience as well as from published research findings.

For some cropping systems, farmers/growers may need to rely on a specialist who is able to monitor and advise on each step in a soil–crop–climatic growing system.

## 27.5  PRECISION FARMING

Probably no other aspect of plant nutrient element management has attracted as much attention as that associated with what has been coined "precision farming," the design of detailed cropland maps based on a combination of grid soil sampling, aerial photography, and on-the-go yield determinations. Using a global positioning system (GPS), site-specific maps can be drawn to correlate soil characteristics (fertility, topography, drainage, etc.) to crop yield. Although detailed soil sampling, referred to as grid sampling (see page 119), is required, grid plant sampling has not been included as a factor in precision farming systems. However, infrared photography can be used to observe plant canopy characteristics that may relate to a particular stress condition. The current state-of-the-art of infrared imaging may not be able to provide meaningful information as to the plant nutrient element status of the photographed crop. However, when changes in canopy characteristics are observed, assaying plants in specific areas can help verify plant nutrient element insufficiencies (see Chapter 17) and, from the assay results, qualifying that imagery that would be identified as associated with a particular plant nutrient element insufficiency.

*Appendices*

# Appendix A: Glossary

These definitions are defined in terms of their application to soil fertility and plant nutrition, terms that may also have wider application and therefore may be differently defined in other scientific disciplines.

## A

**Absorption:** A process in which a substance is taken into something, such as a plant cell or structure by either an active (biological) or passive (physical or chemical) process.

**Accumulator plants:** A plant species that accumulates an element in either its vegetative tissues, fruits, or seeds. Most accumulator elements are either a micronutrient, trace element, or heavy metal. An example of an accumulator plant is the sunflower because Cd accumulates in its seeds.

**Active absorption:** Refers to the process of ion uptake by plant roots requiring the expenditure of energy. This process is controlled and specific as to the numbers and types of ion species absorbed (see passive absorption).

**Adsorption:** The attachment of a substance to the surface of another substance.

**Aerobic:** Having molecular oxygen ($O_2$) as a part of the environment.

**Alkali soil:** A soil that contains substantial quantities of sodium (Na), and whose water pH will be greater than 8.2.

**Alkaline soil:** A soil having a water pH greater than 7.0. A soil that is saturated with calcium carbonate ($CaCO_3$) will have a water pH of 8.2

**Amino acid:** An organic acid containing an amino group ($NH_2$), a carboxyl group (COOH), and an attached alkyl or aryl group.

**Ammonium:** A compound consisting of one nitrogen (N) and four hydrogen (H) atoms that exists as a cation whose formula is $NH_4^+$. The ammonium cation exists in the soil solution and on the soil's cation exchange complex. The ammonium compound is combined with various anions to form N compounds that are used as N fertilizer sources, such as ammonium sulfate [$(NH_4)_2SO_4$], ammonium nitrate ($NH_4NO_3$), diammonium phosphate [$(NH_4)_2HPO_4$] and monoammounium phosphate ($NH_4H_2PO_4$), and anhydrous ammonia ($NH_3$).

**Anaerobic:** An environmental condition in which molecular oxygen ($O_2$) is deficient.

**Anhydrous ammonia:** A compound consisting of one nitrogen (N) and three hydrogen (H) atoms that exists as a gas ($NH_3$) and is used as an N source fertilizer.

**Anion:** An ion in solution carrying a negative charge, examples being borate ($BO_3^{3-}$), chloride ($Cl^-$), dihydrogen phosphate ($H_2PO_4^-$), hydrogen phosphate ($HPO_4^{2-}$), nitrate ($NO_3^-$), nitrite ($NO_2^-$), and sulfate ($SO_4^{2-}$). In chemical notation, the minus sign indicates the number of electrons the compound will give up.

**Anion exchange:** A physiochemical property of some soils in which anions in solution will be physically adsorbed onto colloidal surfaces, holding the anion in place, which can become available for absorption by plant roots.

**Atom:** The smallest unit of a substance that cannot be broken down further or changed to another substance by purely chemical means.

**Atomic weight:** The average mass of a single atom of an element, expressed in terms of a dimensionless unit approximately equal to the mass of one hydrogen atom.

**Atmospheric demand:** The capacity of air surrounding the plant to absorb moisture. This capacity of air will influence the amount of water transpired by the plant through its exposed surfaces. Atmospheric demand varies with changing atmospheric conditions. It is greatest when air temperature and movement are high and relative humidity is low. When the reverse conditions exist, the atmospheric demand is low.

**Availability:** A term used to indicate that an element is in a form and position suitable for plant root absorption.

**Available Plant Water:** That portion of water in the soil that can be readily absorbed by plant roots; that level of water held in the soil between field capacity and the permanent wilting percentage (see page 177).

# B

**Bacteria:** Unicellular organisms that may be pathogens that infect plants, causing disease. Bacteria exist in the soil, performing useful functions that can be beneficial to plants.

**Base:** Any compound that dissociates upon contact with water, releasing hydroxide ions ($OH^-$).

**Beneficial element:** An element not essential for plants but when present for plant use at specific concentrations will enhance plant growth (see Chapter 13).

**Best Management Practices (BMPs):** The use of production practices based on science and proven in the field to be beneficial to plants, affecting growth, fruit production, and quality of product. This concept was introduced with the passage of The Federal Water Pollution Control Act Amendments of 1972, Public Law 92-500, a law that specified practices designed to control non-point water pollution. Best Management Practices, better known by its acronym BMPs, were outlined as a means for effective, practical, structural or nonstructural procedures to prevent the movement of sediment, nutrients, pesticides, and other pollutants from the land to surface or groundwater, thereby protecting water quality (see Chapter 27).

**Best practice:** A technique, method, process, activity, incentive, or reward that is believed to be more effective at delivering a particular outcome than any other technique, method, process, etc. when applied to a particular condition or circumstance (www.Wilipedia.org) (see Chapter 27).

**Boron (B):** An essential element classified as a micronutrient involved in energy transfer and carbohydrate movement in a plant. The element exists in the soil solution as the borate ($BO_3^{3-}$) anion as well as the molecular boric acid

($H_3BO_3$). Common boron (B)-containing fertilizers are solubor ($Na_2B_4O_7 \cdot H_2O + Na_2B_{10}O_{16} \cdot H_2O$) and borax ($Na_2B_4O_7 \cdot 10H_2O$). Other B-containing fertilizers are given in Table 19.1)

**Buffer capacity:** A measure of the ability of a solution to maintain a constant pH by neutralizing excess acid or base.

## C

**C3 plants:** Plant species whose photosynthetic pathway results in the initial formation of a three-carbon-containing carbohydrate. Such plants are most likely to be legumes that reach maximum efficiency at relatively low air temperatures and light intensities, and are highly responsive to high atmospheric carbon dioxide ($CO_2$) concentrations.

**C4 plants:** Plant species whose photosynthetic pathway results in the initial formation of a four-carbon-containing carbohydrate. Such plants are grains and grasses, and are highly responsive to increasing levels of light intensity and less so to increasing carbon dioxide ($CO_2$) content in the atmosphere.

**Calcareous soil:** Soil having a water pH of 8.2 due to the presence of free calcium carbonate ($CaCO_3$) in the soil.

**Calcium (Ca):** An essential plant element classified as a major element that serves as a specific component of organic compounds and exists in the soil solution as the $Ca^{2+}$ cation. Common sources of Ca are various liming materials, calcium oxide (CaO), hydrated lime [$Ca(OH)_2$], calcium carbonate ($CaCO_3$), and the fertilizer calcium nitrate [$Ca(NO_3)_2 \cdot 4H_2O$] (see Chapters 18 and 19).

**Carbon (C):** An essential plant element classed as a major element. Carbon (C) is obtained by the plant from carbon dioxide ($CO_2$) in the air and is incorporated into a carbohydrate during the photosynthesis process in chlorophyll-containing plants (see page 15).

**Catalyst:** A substance whose presence causes or speeds up a chemical reaction between two or more other substances.

**Cation:** An ion having a positive charge, examples being ammonium ($NH_4^+$), calcium ($Ca^{2+}$), copper ($Cu^{2+}$), iron ($Fe^{2+}$ or $Fe^{3+}$), hydrogen ($H^+$), potassium ($K^+$), magnesium ($Mg^{2+}$), manganese ($Mn^{2+}$), and zinc ($Zn^{2+}$). In chemical notation, the plus sign indicates the number of electrons the element will accept.

**Chelate:** A type of chemical compound in which a metallic atom (such as Fe) is firmly combined with a molecule by means of multiple chemical bonds. The term refers to the claw of a crab, illustrative of the way in which the atom is held.

**Chlorine (Cl):** An essential plant element classified as a micronutrient involved in the evolution of oxygen in photosystem II and whose presence raises cell osmotic pressure. Chlorine exists in the plant and soil solution as the chloride ($Cl^-$) anion. A fertilizer source for chlorine (Cl) is potassium chloride (KCl).

**Chlorophyll:** A complex molecule found in green plants that is directly involved in photosynthesis. The chlorophyll molecule is a complex structure with a magnesium (Mg) atom in the center of the ringed structure (see Figure 3.1, page 16).

**Chlorosis:** A light green to yellow coloration of leaves or whole plants that usually indicates an essential element insufficiency or toxicity, frequently related to the essential element Fe.

**Colloid:** A substance (inorganic or organic) that has been subdivided into an extremely small particle capable of forming a colloidal suspension, with the particle size being smaller than 0.001 millimeters (0.00004 inches) in diameter, and which carries a negative charge.

**Compound leaf:** A leaf whose blade is divided into a number of distinct leaflets.

**Copper (Cu):** An essential plant element classified as a micronutrient that participates in electron transport, and protein and carbohydrate metabolism. Copper exists in the plant and soil solution as the cupric cation ($Cu^{2+}$). Common Cu-containing fertilizers are copper sulfate ($CaSO_4 \cdot 5H_2O$), cuprous oxide ($Cu_2O$), and cupric oxide ($CuO$); (see Table 19.1).

**Critical value:** A concentration value of an essential plant element below which deficiency occurs and above which sufficiency exists (see page 135).

## D

**Deficiency:** Describes the condition when an essential plant element is not in sufficient concentration (or form) in the plant to meet the plant's physiological requirement. Plants usually grow poorly and show visual signs of abnormality in color and structure.

**Denitrification:** The process by which soil N [in the nitrate ($NO_3^-$) form] is lost into the atmosphere as dinitrogen oxide ($N_2O$) and molecular nitrogen gas ($N_2$).

$$NO_3^- \rightarrow NO_2^- \rightarrow NO \rightarrow N_2O \text{ (gas)} \rightarrow N_2 \text{ (gas)}$$

**Diagnostic Approach:** An organized procedure applying knowledge and tools to study conditions of plants in order to evaluate the soil/plant condition with the goal of improving the fertility of the soil, or other rooting medium, and the yield and quality of produced plant products.

**Diffusion:** The movement of an ion in solution from an area of high concentration to an area of lower concentration. Movement continues as long as the water films and concentration gradient exist.

**Dry ashing:** A method for destroying the organic constituent in plant tissue in the preparation for elemental analysis (see page 256).

**Drip irrigation (trickle irrigation):** A method whereby water under pressure is applied slowly at a specific point through small orifice emitters.

**DRIS:** An acronym for the Diagnosis and Recommendation Integrated System developed by Beaufils in 1973, it is based on nutrient element relationships comparing an observed value to that of a corresponding ratio (Beverly 1991).

# E

**Electrical conductivity (EC):** A measure of the electrical resistance of water or effluent from the rooting media used to determine the level of ions in solution whose value can be used to determine the potential effect on plant growth; the unit commonly used is decisiemen per meter (dS/m) [dS/m = mS/cm = millimhos per centimeter (mmhos/cm) = 760 ppm].

**Electron:** A tiny charged particle, smallest of the three principal constituents of the atom, with a mass of $9.1 \times 10^{-28}$ grams and a negative charge of $1.6 \times 10^{-19}$ coulomb.

**Element:** A substance made up entirely of the same kind of atom.

**Enzyme:** An organic compound that serves as a catalyst in metabolic reactions. Enzymes are complex proteins that are highly specific to particular reactions.

**Essential elements:** Those elements that are necessary for higher plants to complete their life cycle. Also refers to the requirements established for essentiality by Arnon and Stout in 1939 (see Chapters 11 and 12).

**Evaporation:** The conversion of a substance from a liquid to a gaseous state. Evaporation occurs when a molecule of liquid gains sufficient energy (in the form of heat) to escape from the liquid into the atmosphere above it.

**Evapotranspiration:** The sum of evaporation and transpiration of water from an exposed surface, such as a leaf, with the total sum varying with temperature, wind speed, type of vegetation, leaf characteristics, and relative humidity.

# F

**Fertigation:** Application of fertilizer through an irrigation system.

**Fertilizer:** A substance containing one or more of the essential plant elements that, when added to a soil/plant system, aids plant growth and/or increases productivity by providing additional essential elements for plant use.

**Fertilizer elements:** The three essential major plant elements—nitrogen (N), phosphorus (P), and potassium (K)—are so identified as they are the commonly included elements in a wide range of fertilizer materials.

**Fertilizer grade:** A term used to identify a mixed fertilizer containing two or three of the fertilizer elements in the order, nitrogen (N), phosphorus (P) expressed as its oxide ($P_2O_5$), and potassium expressed as its oxide ($K_2O$), the content of each element given in percent.

**Fibrous root system:** A plant root system consisting of a thick mat containing many short, slender, interwoven roots, without the presence of a tap root.

**Foliar fertilization/feeding:** Application of an essential plant element (or elements) to the foliage of a plant for the purpose of improving the plant's nutritional status.

**Free space:** That portion, about 10%, of plant roots into which ions in the soil solution can passively enter the root without passing through a membrane.

**Fulvic acid:** Organic materials that are soluble in alkaline solution and remain soluble on acidification of the alkaline extracts.

**Fungicide:** A chemical that is applied to a plant to kill disease organisms.

**Fungus:** A microscopic organism with a body of spider web-like filaments, lacking chlorophyll, and having its own ability to manufacture food; many fungi cause plant diseases. Fungi that exist in a soil can alter the soil's biological character, and either increase or decrease the soil's ability to support plant growth.

**G**

**Gram:** A unit of mass in the metric system, equivalent to the mass of one cubic centimeter of water at maximum density (4°C). There are 453.6 grams in an English pound.

**H**

**Half-life:** That time period when half of the original existence of a substance remains. It is a term that is applied to elements or chemicals that have a unique characteristic, such as its biological effect or toxicity.

**Hard pan:** A hard layer of soil beneath the tilled zone through which water and roots cannot easily penetrate. A hardpan may exist naturally or it can be the result of soil manipulation using cultivation tools.

**Heavy metals:** A group of metallic elements with relatively high atomic weights (>55) that generally have similar effects on biological functions. Elements such as cadmium (Cd), chromium (Cr), lead (Pb), and mercury (Hg) are frequently identified as heavy metals. The four micronutrients copper (Cu), iron (Fe), manganese (Mn), and zinc (Zn) have also been frequently referred to also as heavy metals (see Chapter 14).

**Hectare (ha):** A unit of measure; 10,000 $m^2$, equivalent to 2.47 acres.

**Herbicide:** A chemical that kills plants when brought into contact with its foliage or when present in the rooting zone of the plant.

**Host:** A plant that is being attacked by a pathogen.

**Humidity:** Amount of water vapor in the air. Relative humidity is the ratio of the air's actual water content to the total amount of water vapor the same volume of air could theoretically hold under its current conditions of temperature and pressure.

**Humus:** Organic materials within the soil that have been partially decomposed, forming a relatively stable organic matrix that has colloidal properties and, in turn, can have a marked effect on the physical and chemical properties of a soil.

**Humic substances:** Organic materials with chemical structures that do not allow them to be placed into the category of non-humic biomolecules.

**Humic acids:** Organic materials that are soluble in alkaline solution but precipitate on acidification of the alkaline extract.

**Hybrid:** A cross between parents that are genetically different to form a new plant type.

**Hydrocarbon:** Any compound that contains only hydrogen (H) and carbon (C) in its molecular structure.

**Hydrogen (H):** An essential plant element classified as a major element. Hydrogen is obtained from water ($H_2O$); and after splitting the water molecule, the H atom is combined with carbon dioxide ($CO_2$) to form a carbohydrate in the process called photosynthesis (see page 15).

**Hydrogen bond:** An electrostatic bond between an atom of hydrogen (H) in one molecule and an atom of another element in a neighboring molecule.

**Hydroponics:** A method of growing plants without soil in which the essential plant elements are supplied by means of a nutrient solution that periodically bathes the plant roots. The word was coined by the combination of two Greek words, *hydro* meaning water and *ponos* meaning labor, therefore meaning *working water* (see Chapter 23).

**Hydroxide:** Any inorganic compound containing hydrogen (H) and oxygen (O) bound together in the form of a negatively charged ion ($OH^-$).

**Hyphae:** Thread-like filaments forming the mycelium of fungi.

## I

**Ion:** An atom that has lost or gained one or more electrons, and thus acquires either a small positive or negative charge.

**Ion carriers:** A proposed theory that explains how ions are transported from the surface of a root into root cells, the transport being across a membrane and a concentration gradient, the transport occurring by means of a carrier or carrier systems (see Chapter 4).

**Ion exchange:** Refers to the phenomenon of physical–chemical attraction between charged ions, either negative or positive, colloidal substances (such as clay and humus) with cations and anions.

**Ion pumps:** A proposed theory that explains how ions are transported from the surface of a root into root cells, a transport that occurs across a membrane and against a concentration gradient.

**Iron (Fe):** An essential plant element classified as a micronutrient that is involved in the electronic transport systems in plants. Iron can exist as either the ferrous ($Fe^{2+}$) or ferric ($Fe^{3+}$) cation in the plant or soil solution (see Chapter 12).

**Immobilization:** A process occurring in soil in which elements in a soluble form interact with the soil to be rendered into an insoluble form.

**Insecticide:** A chemical substance that kills insects.

## K

**K-Mag:** A naturally occurring source of potassium (21% to 22% $K_2O$), magnesium (10.5% to 11% Mg), and sulfur (21% to 22% S); a highly available, water-soluble sulfate form that can be used as a fertilizer to supply one or all three elements.

**$K_2O$:** A chemical designation for identifying the potassium (K) content in fertilizer, often referred to a "potash."

# L

**Leaching:** Removal of soluble salts by the downward movement of water through the rooting media and/or soil profile.

**Leaf analysis (plant analysis):** A method of determining the total elemental content of a leaf and relating this concentration to the well-being of the plant in terms of its elemental composition (see Chapter 17).

**Leaflet:** The subdivision of a leaf.

**Lignin:** A complex organic compound found as a constituent of the walls of plant cells.

**Liter:** Organic materials derived from plant residues located on the soil surface.

**Luxury consumption:** Absorption of an element by the plant that exceeds that needed for sufficiency. Luxury consumption commonly occurs when plants are being excessively fertilized, the elements N as the $NO_3^-$ anion and K are the two elements that will readily accumulate in the plant.

# M

**Macronutrients:** Refers to the nine essential plant elements—carbon (C), calcium (Ca), hydrogen (H), magnesium (Mg), nitrogen (N), oxygen (O), phosphorus (P), potassium (K), and sulfur (S)—required at relatively high concentrations in the plant (see Chapter 11).

**Magnesium (Mg):** An essential plant element classified as a major element, a constituent of the chlorophyll molecule (see Figure 3.1, page 16), and serves as an enzyme cofactor for phosphorylation processes. Magnesium exists in the plant and soil solution as a divalent Mg ($Mg^{2+}$) cation. Major sources of Mg are dolomitic limestone, magnesium sulfate (epsom salt) ($MgSO_4 \cdot 7H_2O$), magnesium oxide (MgO), and SUL-PO-MG. (see Chapter 11 and 18).

**Major essential elements:** The nine essential plant elements found in relatively large concentrations in plant tissues. These elements are calcium (Ca), carbon (C), hydrogen (H), oxygen (O), magnesium (Mg), nitrogen (N), phosphorus (P), potassium (K), and sulfur (S) (see Chapter 11).

**Manganese (Mn):** An essential plant element classified as a micronutrient that is involved in the oxidation-reduction processes and acts as an enzyme activator. Manganese exists in the plant and soil solution as a cation in several oxidation states ($Mn^{2+}$, $Mn^{3+}$, $Mn^{4+}$). The most common oxidation state is the divalent cation, $Mn^{2+}$. Common fertilizer sources are manganese oxide (MnO) and manganese sulfate ($MnSO_4 \cdot 5H_2O$). (see Table 19.1)

**Mass flow:** The movement of ions in a soil with the flow of water, the ions being carried in the moving water.

**Metabolism:** The set of chemical reactions taking place within living cells that allows energy to be used and transferred, and tissues to be constructed and repaired.

**Micronutrients:** Seven essential plant elements found in relatively small (less than 0.01%) concentrations in plant tissue. These elements are boron (B), chlorine (Cl), copper (Cu), iron (Fe), manganese (Mn), molybdenum (Mo), and zinc (Zn) (see Chapter 12).

**Mineral:** A term sometimes used to identify an essential plant element. The more common word use is to identify substances that exist in a soil, having both chemical and physical properties that determine the character of a soil, and comprise the basic constituents that on weathering (decomposition) release essential plant elements and form the basis for various sized particle forms (sand, silt, and clay).

**Mineral nutrition:** The study of the essential plant elements as they relate to the growth and well-being of plants.

**Mineralization:** The decomposition in soil of both inorganic minerals and organic compounds, releasing soluble forms of elements into the soil solution and leaving residues that form the physical structure of the soil.

**Molecular weight:** The weight of one molecule, ion, group, or other formula unit. The molecular weight is the sum of the atomic weights of the atoms that combine to make up the formula unit.

**Molecule:** The smallest unit into which a compound may be divided and still retain its physical and chemical characteristics.

**Molybdenum (Mo):** An essential plant element classified as a micronutrient. It is a component of two enzyme systems that are involved in the conversion of nitrate to ammonium in a plant. Molybdenum exists in the soil solution as the molybdate ($MoO_3^-$) anion. Ammonium molybdate [$(NH_4)_2Mo_7O_{24} \cdot 2H_2O$], sodium molybdate ($Na_2Mo_4 \cdot 2H_2O$), and molybdenum trioxide ($MoO_3$) (see Chapter 12).

**Mycelium:** A mass of fungal growth consisting of branching, thread-like hyphae.

**Mycorrhizae:** Primarily fungi that are associated with root activity, specific for some plant species, providing both protection as well as participating in the utilization of elements with the rhizosphere.

# N

**Necrosis:** Plant tissue that turns black due to disintegration or death of cells, usually caused by disease.

**Nematode:** Microscopic nonsegmented roundworms that enter the roots of a plant, forming a knot at that point, the result being a stimulation of cellular growth.

**Nickel (Ni):** An element thought to be beneficial, if not essential, to some plants, particularly affecting seed viability. Its available form for root uptake is the divalent $Ni^{2+}$ anion (see Chapter 13).

**Nitrate:** An anion of one atom of nitrogen (N) and three of oxygen (O) to form a monovalent nitrate ($NO_3^-$) anion that is a common form of nitrogen (N) found in soils and that can be readily absorbed by plant roots in order to satisfy the nitrogen (N) requirement of a plant. The nitrate ($NO_3^-$) anion moves in the soil primarily by mass flow.

**Nitrification:** The conversion of ammonium to either nitrite ($NO_2$), nitrate ($NO_3$), or nitrous oxide ($N_2O$) by bacteria that exist in the soil. The reaction is governed by soil temperature (65° to 90°F), aeration (needs oxygen), pH (6.0 to 7.0), and moisture (extreme conditions retard bacterial activity).

$$NH_4^+ \text{ nitromonas bacteria} \rightarrow NO_2^- \text{ Nitrobacter bacteria} \rightarrow NO_3^-$$

$$NO_2^- \rightarrow N_2O$$

**Nitrite:** An anion of one atom of nitrogen (N) and two of oxygen (O) atoms to form the monovalent nitrite ($NO_2^-$) anion, a reduced form of nitrogen (N) that exists mostly under anaerobic soil conditions and is a form of nitrogen (N) that is highly toxic to plants.

**Nitrogen (N):** An essential plant element classified as a major element. It is a component of amino acids and proteins. Nitrogen exists in the atmosphere as a gas ($N_2$), in the plant as proteins and in soil solution as either the nitrate ($NO_3^-$) anion or the ammonium ($NH_4^+$) cation. Nitrogen is classified as a fertilizer element and is the first element given a fertilizer grade designation, expressed as a percent. The chemical form of nitrogen (N) in the fertilizer may be identified specifically as to its compound form and/or as either in the ammonium or nitrate form. See Table 19.1 for common nitrogen (N) fertilizers.

**Nitrogen-fixing bacteria:** Any of several genera of bacteria that have the ability to convert atmospheric nitrogen ($N_2$) into the ammonium ($NH_4$) and nitrate ($NO_3$) forms of N, which can then be used by other organisms.

**Nutrient:** In plant nutrition use, refers to one of the essential plant elements and is frequently used synonymously with the word "element." A nutrient may also designate a substance that affects the growth and development of a plant.

**Non-point source pollution:** Those pollutants with diffuse sources, such as runoff from farms, forests, pastures, construction sites, strip mines, urban lawns and streets, and rural sanitation facilities.

**O**

**Organic farming/gardening:** Defines a method of crop production that is conducted without the use of man-made chemicals to provide essential plant elements, control plant growth, or protect plants from diseases and insects. Not being a specific term, what would be defined as "organic" can be found in state and/or federal regulations or laws (see Chapter 25).

**Organic matter:** Material that was formed by the bodily processes of an organism and as plant residues. Decaying organic matter (plant and organism residues) can be a significant source of several essential plant elements, primarily nitrogen (N), phosphorus (P), sulfur (S), and boron (B) (see Chapter 10).

**Osmosis:** The passage of water or another solvent through a membrane in response to differences in the concentrations of dissolved materials on opposite sides of the membrane.

**Osmotic pressure:** Force exerted by substances dissolved in water that affects water movement into and out of plant cells. Salts in the soil solution exert some degree of force that can restrict water movement into plant root cells or extract water from them.

**Oxygen ($O_2$):** An essential plant element classified as a major element. Oxygen is combined with hydrogen (H) and carbon (C) in the photosynthetic process to form carbohydrates that form the plant structure (see page 16).

**Ozone ($O_3$):** An unstable, faintly bluish gas; the most chemically active form of oxygen ($O_2$). Ozone can be used as a disinfectant. When formed in the atmosphere as a result of various forms of combustion, its presence in the atmosphere around plants can cause plant injury.

**P**

**Passive absorption:** The movement of ions into plant roots carried along with water being absorbed by roots.

**Pesticide:** A general term that describes a chemical or natural substance used to control any sort of plant pest.

**Petiole:** The stem of a leaf that connects the base of the leaf blade to the stem. The petiole can be a useful plant part for determining the essential element status of a plant, nitrate ($NO_3$) being the primary analyte assayed. Some elements, such as potassium (K) and nitrogen (N), accumulate in leaf petioles.

**pH:** A measure of the acidity or basicity of a liquid, a measure of the number of hydrogen ($H^+$) ions present in the liquid (in moles/L), expressed as the negative logarithm. pH ranges from 1 (acid) to 14 (alkaline) (see Chapter 9).

**Phloem:** Vascular plant tissue system through which sugar and other food substances formed in the leaves are conducted to other parts of the plant. The direction of flow is downward, the driving force being osmotic pressure.

**Phosphorus (P):** An essential plant element classified as a major element. It is a component of several enzymes and proteins and an element involved in various energy transfer systems in the plant. Phosphorus exists in the soil solution as an anion in various forms, monohydrogen phosphate ($HPO_4^{2-}$) or dihydrogen phosphate ($H_2PO_4^-$) depending on the pH. Phosphorus is classified as a fertilizer element, being the second ingredient given in percent content as the phosphorus pentoxide ($P_2O_5$) form. See Table 19.1 for a list of commonly use phosphorus (P)-containing fertilizers.

**Photoperiodism:** A term describing the behavior of a plant in relation to the hours of daylight or darkness to which it is exposed.

**Photosynthesis:** The synthesis of carbohydrates from carbon dioxide ($CO_2$) and water ($H_2O$) by living organisms utilizing light energy as an energy source; is catalyzed by chlorophyll according to the following formula: $6CO_2 + 6H_2O$ in light $\rightarrow C_6H_{12}O_6 + 6O_2$ (see page 16).

**Photosynthetically active radiation (PAR):** That portion of the spectrum of solar radiation (450 to 700 nm) that participates in photosynthesis.

**Physiological disorder:** A disorder not caused by pathogens, but instead due to some physiological dysfunction of the organism.

**Plant analysis (leaf analysis):** A method of determining the total elemental content of the whole plant or one of its parts, and then relating the concentration found to the well-being of the plant in terms of its elemental requirement (see Chapter 17).

**Plant canopy:** Refers to the grouping of plants into a unit that has its own internal environmental characteristics that are frequently different from the surrounding atmospheric environment.

**Plant nutrients:** A term used to identify those elements that are essential to plants (see Chapter 3). Sometimes the term used will be "plant nutrient element."

**Plant nutrition:** The study of the effects of the essential as well as other elements on the growth and well-being of plants.

**Plant requirement:** That quantity of an essential element needed for the normal growth and development of the plant without inducing stress due to its deficiency or excess.

**Point source pollution:** A polluting substance that emanates from a pipe or direct and specific point.

**Potash:** A designation for identifying the potassium (K) content in fertilizer, expressed as the oxide of potassium, $K_2O$.

**Potassium (K):** An essential plant element classified as a major element. Its primary function is to maintain cellular turgor. Potassium exists in the plant and soil solution as a monovalent potassium cation ($K^+$). Potassium is classified as a fertilizer element and is the third ingredient given as the oxide form ($K_2O$) in percent. See Table 19.1 for a list of commonly used potassium (K)-containing fertilizers.

**ppb:** An abbreviation for parts per billion, which is 1/1000 of a part per million, expressed in metric units as micrograms per kilogram ($\mu g/kg$) in weight units and in micrograms per liter ($\mu g/L$) in liquid units.

**ppm:** An abbreviation for parts per million, expressed in metric units as milligrams per kilogram (mg/kg) in weight units and in milligrams per liter (mg/L) in liquid units.

**Precision agriculture:** A method for scheduling and carrying out procedures for the establishment and management of a crop production system.

**R**

**Relative humidity:** The ratio of the actual amount of water in the air to the maximum amount (saturation) that the air can hold at the same temperature, expressed as a percentage between 0 and 100.

**Respiration:** A set of processes through which energy is obtained from the decomposition of sugars and other carbohydrates existing in the body of a living organism.

**Rhizosphere:** That thin cylindrical column surrounding the root that is acidic in nature and an area of physical and biological activity, reflecting the physical, chemical, and biological aspects of the plant root and surrounding microflora (see Chapter 4).

**Root:** The part of the plant, usually underground, whose main functions are to anchor the plant, serving as the conduit for the absorption of water and elements as ions into the plant (see Chapter 4).

**Root interception:** Expansion (growth) of plant roots into a rooting medium that increases the contact area between root and soil surfaces, thereby enhancing the absorption of both water and elements into the roots.

**S**

**Saline soil:** Contains soluble salts in such quantities that they interfere with plant growth. The primary element in excess is usually sodium as its $Na^+$ cation, although it can be other elements as their ions.

**Scorch:** Burned leaf margins, a visual symptom typical of potassium (K) deficiency or boron (B) and chloride (Cl) excess (see Table 3.6).

**Scouting:** A term describing an orderly procedure for walking a crop field looking for insects, their number, and the species present.

**Secondary elements:** An obsolete term once used to identify the essential plant elements, calcium (Ca), magnesium (Mg), and sulfur (S), which are now classified as major essential nutrient elements (see Chapter 11).

**Senescence:** A physiological aging process in which tissues of an organism deteriorate and finally die.

**Silicon (Si):** An element believed to be beneficial to plants, particularly small grains (mainly rice – for stem strength), and for other plants providing some degree of fungal disease resistance. The available form of silicon (Si) is the divalent silicate ($SiO_3^{2-}$) anion (see Chapter 13 and page 199).

**Slow-release fertilizer:** A form of fertilizer that has been chemically treated or coated so that its solubility can be controlled when applied to the rooting medium; availability determined by moisture and temperature conditions.

**Sodic soil:** Contains sufficient sodium (Na) adsorbed on the clay particles to interfere with plant growth.

**Sodium absorption ratio (SAR):** A ratio used to express the relative activity of the sodium cation ($Na^+$) in relation to the calcium ($Ca^{2+}$) and magnesium ($Mg^{2+}$) cations, expressed in milliequivalents per liter (meq/L); $SAR = Na^+ [Ca^{2+} + Mg^{2+})/2]^{1/2}$. SAR is used to define a soil condition that would significantly and adversely affect a plant.

**Soil organic matter:** The sum total of all natural and thermally altered, biologically derived organic material found in the soil or on soil surfaces irrespective of the source, whether it is living or dead, or stage of decomposition, but excluding the aboveground portions of living plants (see Chapter 10).

**Soil separates:** Three soil mineral particles defined by their physical size, sand (2.0 to 0.05 mm), silt (0.05 to 0.002 mm), and clay (<0.002 mm). The relative ratio of the separates will define the soil's textural class (see page 50).

**Soil solution:** The liquid, water phase of the soil that contains solutes and exists as water films around soil particles.

**Soil testing:** A method of collecting and assay a soil to determine its fertility status and, based on the assay results, formulate a lime and fertilizer recommendation (see Chapter 16).

**Soil texture:** One of twelve soil textural classes (clay, sandy clay, silty clay, clay loam, sandy clay loam, silty clay loam, loam. sandy loam, silt loam, loamy sand, silt, clay) defined by the percentage of sand, silt, and clay present in a soil. Soil texture plays a major role in defining the physical and chemical properties of a soil (see Chapter 7).

**Soilless culture:** A term used to describe the growing of plants in substances other than soil (see Chapters 23 and 24).

**Soluble salts:** A term that defines the concentration of ions in water (or nutrient solution) measured in terms of its electrical conductivity; used to determine the quality of water, the soil solution, and a hydroponic nutrient solution.

**Specific conductance:** The reciprocal of the electrical resistance of a solution, measured using a standard cell and expressed as decisiemens per meter (dS/m) at 25°F.

**Standard value:** The mean concentration of an essential element in plant tissue based on the analysis obtained from a large population of normal growing plants (see page 136).

**Structural elements:** The essential elements, carbon (C), hydrogen (H), and oxygen (O), that constitute 90% of the dry weight of a plant, and are combined in the photosynthesis process to form carbohydrates (see page 16).

**Subsoil:** That portion of the soil profile that is just below the surface soil. The subsoil normally has physical and chemical properties distinctly different from those of the surface soil, and it may be defined based on the depth of the surface soil, which mainly constitutes the rooting zone.

**Sufficiency and sufficiency range:** That concentration of an essential plant element required to satisfy the plant's physiological requirement for normal, healthy growth leading to high quality production (see page 136).

**Sulfur (S):** An essential plant element classified as a major element. It is a component of some proteins and is a component of glucosides that are the source of the characteristic odors of some plants. Sulfur exists in the plant and soil solution as the divalent sulfate anion ($SO_4^{2-}$). A list of the commonly used sulfur (S)-containing fertilizers are listed in Table 19.1.

**Surface soil:** That portion of the soil profile where the majority of plant roots exist. The surface soil may be of natural depth or that created by tillage operations (see Topsoil).

**Symbiotic bacteria:** Relates to bacteria that infect plant roots of legumes forming nodules on the roots, fixing atmospheric nitrogen ($N_2$), thereby providing fixed nitrogen (N) for the plant while obtaining their carbohydrate food source from the host plant. Most symbiotic bacteria require cobalt (Co) in order to function.

**T**

**Tap root:** The main descending root of a plant. Normally, plants that have a tap root do not also have a fibrous root system.

**Tissue testing:** A method for determining the concentration of the soluble form of an essential element in the plant by analyzing sap that has been physically extracted from a particular plant part, usually from stems or petioles. Tests are usually limited to the determination of nitrate, phosphate, potassium, and iron (see Chapter 17).

**Topsoil:** That portion of the soil profile that exists from the surface to that depth where there begins a significant change in the physical and/or chemical properties of the soil, and that portion of the soil where the major amount of plant roots exist (see Surface soil).

**Total dissolved solids (TDS):** The concentration of ions in solution measured in milligrams per liter (mg/L) or parts per million (ppm). TDS is related to the EC: EC (in dS/m) $\times$ 640 = TDS (mg/L or ppm).

**Toxicity:** The ability of a substance or element to disrupt the normal functions of a plant.

**Toxicity symptoms:** Visual symptoms of toxicity may not always be the direct effect of the element in excess on the plant, but rather the effect of the excess element on one or more other elements. For example, an excessive level of K in a plant can result in either a Mg and/or Ca deficiency, excess P can result in a Zn deficiency, and excess Zn in an Fe deficiency.

**Trace elements:** An obsolete term that was once used to identify the micronutrients, but today is used to identify those nonessential elements found in plant tissues at low concentrations (less than 1 ppm), and are not essential plant elements, but their presence in the plant may affect the physiology of the plant, or whose present in harvested portions of the plant may pose a hazard when consumed by animals and man (see Chapter 15).

**Tracking:** A technique of following through time the essential plant element content of the rooting media or plant by a sequence of analyses made at specified stages of plant growth. Tracking can be used as a means of determining treatments needed to maintain the soil and plant within their sufficiency range.

**Transpiration:** The process of water loss from plant tissues, usually through the leaves, into the atmosphere.

**Turgidity:** The state of the plant's water content. A turgid plant is one that is stiff and upright due to high internal water pressure; the opposite condition is wilting due to the lack of water and low internal water pressure.

**Turgor:** Normal inflation of cells due to internal water pressure against the cell walls.

**V**

**Valence:** The combining capacity of atoms or groups of atoms defines their valance. For example, the element potassium ($K^+$) and the multiple-atom compound ammonium ($NH_4^+$) are monovalent, whereas calcium ($Ca^{2+}$) and magnesium ($Mg^{2+}$) are divalent. Some elements can have more than one valance state; for example, ferrous iron ($Fe^{2+}$) is divalent, while ferric iron ($Fe^{3+}$) is trivalent.

**Variety (botanical):** A subdivision of a species with distinct morphological characteristics given a Latin binomial name according to the rules of the International Code of Botanical Nomenclature.

**Virus:** Microscopic organism having a strand of nucleic acid surrounded by a protein coat and capable of causing diseases in plants.

**W**

**Water:** Dihydrogen oxide ($H_2O$). Among the most familiar and ubiquitous of all chemical substances, water is also among the most unusual, with a group of unique properties that places it in a class by itself. Water is highly polar, can exist in all three phases of matter—liquid, solid, and gaseous, has an extremely high surface tension, and is considered a universal solvent (see Chapter 22).

**Wet acid digestion:** A method for destroying the organic component in plant tissue using nitric acid ($HNO_3$) plus an oxidizing agent such as 50% hydrogen peroxide ($H_2O_2$) or perchloric acid ($HClO_4$), to determine the elemental content in plant tissue (see page 256).

**Wilt:** A condition in which the plant droops because of a decrease in cell turgor due to the lack of water. The opposite of turgor.

**Wilting percentage:** That level of soil moisture content that results in plant wilting (see page 52).

**X**

**Xylem:** A vascular system in the plant through which water and dissolved substances are transported upward from the roots into the stems, leaves, and fruit.

**Z**

**Zinc (Zn):** An essential plant element classified as a micronutrient. It is involved in several enzymatic functions in plants. Zinc exists in the soil solution as a divalent cation ($Zn^{2+}$). Common Zn-containing fertilizers are given in Table 19.1.

# Appendix B: Formulation and Use of Soil Extraction Reagents

The concept that elemental soil availability could be determined by extracting a soil with a specially formulated reagent had its beginnings in the early 1930s. The quantity of element extracted, however, was found to have no value in and of itself. It was the degree of correlation between quantity of element extracted and plant growth that formed the basis for acceptance of an extraction reagent and related procedures. Those early engaged in this research area were soil scientists who understood the chemical characteristics of soils needed for formulating an extraction reagent and associated application (soil type and characteristics) and use procedures (soil extractant ratio and extraction time).

## B.1  HISTORICAL BACKGROUND

Soil test methodology development occurred in the research laboratories of state land-grant colleges and universities, mainly in the United States, with the application of these methods being promoted by their associated cooperative extension services. Soil testing laboratories were established within these institutions to provide the analytical service needed by farmers and growers to utilize the soil testing methodology for developing lime and fertilizer recommendations necessary to achieve plant nutrient element sufficiency.

A number of national fertilizer manufacturers established soil testing laboratories to provide customer service for those using their fertilizer products, followed shortly thereafter by commercial-regional soil testing laboratories. With this commercialization of soil testing, there came the need to standardize soil testing methods and to adopt a uniform system for determining and expressing a soil test assay result. The Council on Soil Testing and Plant Analysis (currently known as the Soil and Plant Analysis Council) came into existence in 1976, and shortly thereafter published the first edition of the *Reference Methods for Soil Testing,* which in following years, went through various editions; the last in this series was published in 1992 (Anonymous, 1992). Regional soil testing committees consisting of university research and extension personnel, plus those in commercial soil testing businesses, published soil testing manuals and conducted research on the use of soil testing methods for formulating lime and fertilizer recommendations. This historical summary can be found in the books by Carter (1993) and Jones (2001).

Along with the development of soil testing procedures came analytical advancements, making the laboratory assay procedures faster and more reliable. Multi-element analyzers favored the development of multi-element extraction procedures. Jones (2001) has described the principles of elemental analysis associated with soil extractant assay procedures.

Today, most of the soil tests conducted for farmers and growers are about equally divided between state and commercial laboratories, the division ratio depending on the state and region.

## B.2  EXTRACTION REAGENTS

Extraction reagents were developed on the basis of what element or elements could be extracted, suitability based on the characteristics of the soil itself, pH, texture, and organic matter content. Most extraction reagents are identified by the individual who developed them.

Single element extraction reagents for P, Bray P1 and P2 (Bray and Kurtz, 1945) and Olsen P (Olsen et al., 1954), are still in wide use today.

The first multi-element extraction reagent was the Morgan Extractant (Morgan, 1941), identified as being "universal," as determined "soil availability" could be determined for the elements, P, K, Ca, and Mg. The neutral normal ammonium acetate extractant for the determination of the "exchangeable cations $K^+$, $Ca^{2+}$, and $Mg^{2+}$ was initially used to calculate the cation exchange capacity (CEC, see page 55) of a soil based on the addition of the amount of exchangeable cations combined with a determination of the exchangeable $H^+$ cation (Schollenberger and Simon, 1945). And then the two Mehlich extractants: Mehlich No. 1 (Mehlich, 1953), known in the past as either the "North Carolina extractant," or "double acid extractant," for application to the acid sandy soils of the eastern Coastal Plain; and Mehlich No. 3 (Mehlich, 1984) for use with most acid soils.

For alkaline soils, a modification of the Olsen P extractant is being used— ammonium bicarbonate-DTPA, which includes P as well as the K and Mg cations (Soltanpour and Workman, 1979). Current extraction reagents in common use, date when first published, and soil adaptation are as follows:

| Extraction Reagent | Date[a] | Adapted Range of Soil Properties |
|---|---|---|
| Bray P1 | 1945 | Acid soils |
| Bray P2 | 1945 | Acid soils containing calcium phosphate |
| Morgan | 1931 | Acid soils |
| Ammonium acetate | 1945 | Acid to slightly alkaline soils |
| Mehlich No. 1 | 1953 | Acid (<6.5) coastal plain sandy soils of low CEC (<10 meq/100g) and low (<5%) organic matter |
| Olsen | 1954 | Alkaline soils |
| AB-DTPA | 1977 | Calcareous, alkaline, or neutral pH soils |
| Mehlich No. 3 | 1984 | Most acid soils |

[a] Date when method first described.

There are other extraction reagents in use that have been devised to meet particular soil-crop conditions, most adapted by some state-operated soil testing laboratories and not widely used outside a particular state or soil-climatic region.

Recently, equilibrium solutions for evaluating elemental soil availability of the essential plant nutrient elements have come into use, 0.01M $CaCl_2$ being one of them (Houba et al., 2000).

Through years of wide use and correlation research, some of these extraction reagents are also able to assay for elements other than those initially proposed, primarily some of the micronutrients. Such inclusions can have limited application based on the reliability of the correlation between quantity of element extracted and plant growth. Extraction reagents and elements determined include the following:

| Extraction Reagent | Element(s) Determined |
| --- | --- |
| Bray P1 | P |
| Bray P2 | P |
| Olsen | P |
| Morgan[a] | P. K. Ca, Mg |
| Neutral normal ammonium acetate | K, Ca, Mg, Na |
| Mehlich No. 1 | P, K, Ca, Mg, B, Zn |
| Mehlich No. 3 | P, K, Ca, Mg, B, Zn |
| AB-DTPA | P, K, Mg, Cu, Mn, Fe, Zn |

[a] A modification of the Morgan extractant by Wolf (1982) is able to determine the micronutrients, B, Cu, Fe, Mn, and Zn

## B.3   SOIL SAMPLE PREPARATION

Before an extraction can be conducted, the field-collected soil sample must be air-dried, gently crushed, and then passed through a 2-mm screen. That portion passing through the screen is recovered, becoming the sample ready for extraction.

## B.4   EXTRACTION REAGENT FORMULATIONS AND USE

The formulation and use procedures for the commonly used extraction reagents are as follows:

1. Bray P1
   - **Formulation**: Pipette 30 mL 1N $NH_4F$ [weigh 37 g ammonium fluoride ($NH_4F$) into a 1-L volumetric flask and bring to 1,000 mL with pure water] with 50 mL of 0.5 NHCl [pipette 20.4 mL conc. hydrochloric acid (HCl) to 500 mL pure water] in a 1-L volumetric flask and dilute to the mark with pure water.
   - **Use Procedures**: Weigh 2.0 g or scoop 1.70 $cm^3$ air-dried soil <10-mesh-sieved (2-mm) soil into a 500-mL flask and pipette 20 mL Bray P1 extractant into the flask. Shake for 5 minutes and immediately filter through a Whatman No. 2 filter paper. Save the filtrate for P concentration determination.

2. Bray P2
   - **Formulation**: Pipette 30 mL 1N $NH_4F$ [weigh 37 g ammonium fluoride ($NH_4F$) into a 1-L volumetric flask and bring to 1,000 mL with pure water] with 200 mL of 0.5 NHCl [pipette 20.4 mL conc. hydrochloric acid (HCl) to 500 mL pure water] in a 1-L volumetric flask and dilute to the mark with pure water.
   - **Use Procedures**: Weigh 2.0 g or scoop 1.70 $cm^3$ air-dried soil <10-mesh-sieved (2-mm) soil into a 500-mL flask and pipette 20 mL Bray P1 extractant into the flask. Shake for 5 minutes and immediately filter through a Whatman No. 2 filter paper. Save the filtrate for P concentration determination.

3. Olsen P
   - **Formulation**: Weigh 42.0 g sodium bicarbonate ($NaHCO_3$) into a 1-L volumetric flask and bring to 1,000 mL with pure water. Adjust the pH to 8.5 using either 50% sodium hydroxide (NaOH) or 0.5N hydrochloric acid (HCl). Add several drops of mineral oil to avoid exposure of the solution to air.
   - **Use Procedures**: Weigh 2.5 g or scoop 2 $cm^3$ air-dried soil <10-mesh-sieved (2-mm) soil into a 250-mL flask and pipette 50 mL Olsen extractant into the flask. Shake for 30 minutes and immediately filter through a Whatman No. 2 filter paper. Save the filtrate for P concentration determination.

4. Morgan (~0.5N $Na_2C_2H_3O_2$, buffered at pH 4.8)
   - **Formulation:** Weigh 100 g sodium acetate ($Na_2C_2H_3O_2•3H_2O$) in about 900 mL of water in a 1000-mL volumetric flask. Add 30 mL glacial acetic acid ($H_2C_2H_3O_2$). Adjust the pH to 4.8 and dilute to 1000 mL with water.
   - **Use Procedures:** Scoop 5 $cm^3$ air-dried soil <10-mesh-sieved (2-mm) soil into a 50-mL flask and pipette 25 mL Morgan extractant into the flask. Shake for 5 minutes and immediately filter through a Whatman No. 2 filter paper. Save the filtrate for P, K, Ca, and Mg concentration determination.

5. Neutral normal ammonium acetate (1N $NH_4C_2H_3O_2$, pH 7.0)
   - **Formulation**: Weigh 77.1 g ammonium acetate ($NH_4C_2H_3O_2$) into a 1-L volumetric flask and bring to about 900 mL with pure water. Adjust the pH to 7.0 by adding either 3N acetic acid ($CH_3COOH$) or 3N ammonium hydroxide ($NH_4OH$), and then dilute to the mark with pure water.
   - **Use Procedures**: Weigh 2.0 g or scoop 1.70 $cm^3$ air-dried soil <10-mesh-sieved (2-mm) soil into a 500-mL flask and pipette 20 mL BrayP1 extractant into the flask. Shake for 5 minutes and immediately filter through a Whatman No. 2 filter paper. Save the filtrate for K, Ca, and Mg concentration determination.

6. Mehlich No. 1 (formally known as double acid or North Carolina extractant)
   - **Formulation**: Pipette 4 mL conc. hydrochloric acid (HCl) into a 1-L volumetric flask, add 0.7 mL conc. sulfuric acid ($H_2SO_4$), and then dilute to the mark with pure water.
   - **Use Procedures**: Weigh 5 g or scoop 4 $cm^3$ air-dried soil <10-mesh-sieved (2-mm) soil into a 500-mL flask and pipette 15 mL Mehlich No.

1 extractant into the flask. Shake for 5 minutes and immediately filter through a Whatman No. 2 filter paper. Save the filtrate for P, K, Ca, and Mg concentration determination.

7. Mehlich No. 3
- **Formulation**: Weigh 138.9 g ammonium fluoride ($NH_4F$) into a 1-L volumetric flask and add 900 mL pure water. Weigh into this mixture 73.05 g ethylenediaminetetraacetic acid (EDTA) and shake to dissolve, and then dilute to the mark with pure water ($NH_4F$-EDTA Stock Solution). In a 4,000-mL volumetric flask, add 3,000 mL pure water, and weigh 80 g ammonium nitrate ($NH_4NO_3$) into the flask, and then pipette 16 mL of the $NH_4F$-EDTA solution as prepared above, 46 mL glacial acetic acid ($CH_3COOH$), followed by 3.26 mL concentrated hydrochloric acid (HCl).
- **Use Procedures**: Scoop 5 cm$^3$ air-dried soil <10-mesh-sieved (2-mm) soil into a 250-mL flask and pipette 50 mL Mehlich No. 3 extractant into the flask. Shake for 5 minutes and immediately filter through a Whatman No. 2 filter paper. Save the filtrate for P, K, Ca, Mg, and micronutrient concentration determination.

8. Ammonium bicarbonate-DTPA
- **Formulation**: Weigh 9.85 g diethylenetriaminepentaacetic acid (DTPA) into 1,000 mL of pure water [shake for 5 hours to dissolve]. Measure 900 mL of the DTPA solution into a 1-L volumetric flask. Weigh 79.06 g ammonium bicarbonate ($NH_4HCO_3$) and add gradually and with gentle stirring to the 900-mL DTPA solution. After it has dissolved, dilute to the mark with pure water. Adjust the pH to 7.6 with dropwise additions of 2M hydrochloric acid (HCl). Add several drops of mineral oil to prevent exposure to air.
- **Use Procedures**: Weigh 10 g air-dried soil <10-mesh-sieved (2-mm) soil into a 125-mL flask and pipette 20 mL ammonium bicarbonate-DTPA extractant into the flask. Shake for 15 minutes and immediately filter through a Whatman No. 42 filter paper. Save the filtrate for P, K, and micronutrient concentration determination.

9. 0.01M calcium chloride
- **Formulation**: Weigh 1.47 g calcium chloride ($CaCl_2 \cdot 2H_2O$) into a 1-L volumetric flask and dilute to the mark with pure water.
- **Use Procedures**: Weigh 10 g air-dried soil <10-mesh-sieved (2-mm) soil into a 250-mL flask and pipette 100 mL $CaCl_2$ extractant at 20°C into the flask. Shake for 2 hours at 20°C and immediately filter through a Whatman No. 42 filter paper. Save the filtrate for elemental concentration determination.

## B.5  EXTRACTION PROCEDURES FOR THE MICRONUTRIENTS

Recently, considerable research has been devoted to adapting the currently used extraction reagents for micronutrient determinations. The value of the determined extracted amount depends on various factors, such as soil type and plant species, meaning that the interpretation of the extracted value obtained must be compared against those soil and plant factors that are related. Most micronutrient soil tests are of limited use as they must correlate with those soil conditions (see page 122) and plant factors (see page 84) associated with that micronutrient.

# Appendix C: Preparation Procedures and Elemental Content Determination for Plant Tissue

A plant tissue sample is prepared for elemental content determination by a set of procedures that have been standardized and are followed by most plant analysis laboratories.

## C.1 PLANT TISSUE PREPARATION PROCEDURES

The following is the sequence of procedures

1. Moisture removal
2. Particle size reduction
3. Organic matter destruction

### C.1.1 MOISTURE REMOVAL

Plant tissue is oven dried at 176°F (80°C), the required time for complete water removal depending on the initial water content of the tissue, the type of tissue, and the type of drying oven used. To avoid "cooking" the tissue, the sample is loosely placed in a clean paper bag or spaced on a drying sheet. Water removal for tissue high in sugar content is best done by freeze-drying. Once the tissue is dried, store it in a dry environment. Drying at a higher temperature can result in dry matter loss, and at a lower temperature, incomplete water removal. Both conditions can affect an assay result due to the difference in dry matter content.

### C.1.2 PARTICLE SIZE REDUCTION

Particle size reduction is done by either crushing or milling in order to homogenize the tissue sample. Crushing can be done by hand using an agate mortar and pestle, or ground (milled) mechanically using a cutting (Wiley) or abrasion (Tecator) mill fitted with either a 20- or 40-mesh screen. The fineness of particle size reduction is determined in part by the aliquot size needed for a particular assay procedure. Milling can segregate the tissue into component fractions, particularly when finely ground; therefore, careful mixing when measuring an aliquot is required to ensure homogeneity.

Contamination can occur, depending on the composition of the milling or abrasion material. If, for example, the particle size reduction device has brass fittings,

Cu and Zn will be added to the sample and if the device is made of tool or stainless steel, Fe will be added to the sample. For some abrasion milling devices, either Al or Na can be added to the sample. The extent of these additions will depend on the length of contact time and the physical condition of the milling device. The extent of contamination can be estimated by comparing assay results between that prepared by agate crushing versus milling or the abrasion procedure.

### C.1.3    ORGANIC MATTER DESTRUCTION

The organic matter in the tissue can be destroyed by either of two procedures: wet acid oxidation (digestion) or high-temperature oxidation.

- *Wet acid digestion:* A tissue sample aliquot is digested at 100°C in concentrated nitric acid ($HNO_3$) either alone or with an added oxidant, such as 50% hydrogen peroxide ($H_2O_2$) or perchloric acid ($HClO_4$). The use of $HClO_4$ requires the use of a specialized fume hood. Wet acid digestion can be carried out in either a covered beaker, digestion tube, or flask. Wet acid digestion can be done under pressure in a microwave apparatus. Wet acid digestion may result in the loss of B depending on the Ca tissue content, high Ca acting a trap reducing the potential for B loss. Incomplete digestion can occur if the oxidant is ineffective in the dissolution of all the tissue components.
- *Dry ashing (combustion):* A tissue aliquot is weighed into a porcelain or quartz crucible, which is then placed into a cool muffle furnace and the temperature allowed to ramp up to the 500°C temperature slowly so that the tissue does not catch fire. Dry ashing at a lower temperature may result in incomplete oxidation; at higher temperatures, the loss of some elements; and for a shorter time than 4 hours, incomplete oxidation. Dry ashing for longer periods does not affect the oxidation process. Placing large numbers of tissue samples in the muffle furnace may require longer ashing times and the passing of air into the furnace so that sufficient oxygen ($O_2$) is supplied to complete the oxidation process. High carbonaceous or fibrous tissues require longer periods for complete oxidation to take place. Some tissue types, those high in fiber, cellulose, or sugars, may require the use of an ashing aid to achieve complete combustion. One such aid is wetting the plant tissue with a 10% solution of magnesium nitrate [$Mg(NO_3)_2 \cdot 4H_2O$] and drying the tissue in a 100°C oven. This procedure usually ensures complete oxidation of the organic fraction and also prevents the loss of S.

Solubilization of the ash is either with the use of dilute nitric acid ($HNO_3$) or a mixture of $HNO_3$ and hydrochloric acid (HCl). The concentration of the acids required to solubilize the ash depends on the initial sample size and the requirements of the procedures to be used for determining the element content of the acidified ash solution.

## C.2   ELEMENTAL CONTENT DETERMINATIONS

A number of different analytical procedures can be used for determining the essential plant nutrient elements, both the major elements and micronutrients, in the prepared digest or dissolved plant ash. There have been significant advances made in analytical procedures that can be applied for the determination of these elements. Procedures in common use in plant analysis laboratories include

- Dry combustion or Kjeldahl digestion for N
- Dry combustion, turbidic spectrophotometry, or ICP for S
- Spectrophotometry for B, P, and Mo
- Flame emission spectrophotometry for K and Na
- Flame atomic absorption spectrophotometry for Ca, Cu, Fe, Mg, Mn, and Zn
- Flameless atomic absorption spectrophotometry for Mo
- Specific-ion electrodes or ion chromatography for the determination of chloride ($Cl^-$), sulfate ($SO_4^{2-}$), and nitrate ($NO_3^-$) anions

More recently, inductively coupled plasma emission spectrophotometry, frequently known by its acronym ICP or ICAP, has become the analytical procedure of choice because most of the elements determined by either spectrophotometry and flame emission or atomic absorption spectrophotometry are easily assayed by the ICP technique simultaneously. Laboratories using this analytical methodology are able to provide rapid, low-cost plant analysis service.

## C.3   TISSUE TESTING EXTRACTION PROCEDURES

Tissue testing primary relates to the extraction of nitrate nitrogen ($NO_3$-N) as well as P and K from dried ground plant tissue, using conductive tissue such as plant stems or leaf petioles. The various extraction reagents are dilute hydrochloric acid (HCl), 2% acetic acid ($CH_3COOH$), or water, with or without the addition of activated charcoal in order to obtain a clear solution after extraction and filtering. For the determination of $NO_3$-N, 0.025M aluminum sulfate [$Al(SO_4)_3$] is the extraction reagent recommended by some.

# Appendix D: Weights and Measures

## METRIC CONVERSION CHART

1 pound (lb) = 454 grams (g)
1 kilogram (kg) = 2.205 pounds (lb)
1 gallon (gal) = 3.785 liters (L)
1,000 grams (g) = 1 kilogram (kg)
1,000 milliliters (mL) = 1 liter (L)

## EQUIVALENT GFS PACKAGE SIZES (NOT EXACT CONVERSIONS)

| Weights | Volumes |
|---|---|
| 25 grams (g) = 1 ounces (oz) | 25 milliliters (ml) = 1 ounce (oz) |
| 100 grams (g) = 4 ounces (oz) | 50 milliliters (ml) = 2 ounces (oz) |
| 250 grams (g) = 8 ounces (oz) | 125 milliliters (ml) = 4 ounces (oz) |
| 500 grams (g) = 1 pound (lb) | 250 milliliters (ml) = 8 ounces (oz) |
| 2.5 kilograms (kg) = 5 pounds (lb) | 500 milliliters (ml) = 1 pint (pt) |
| 5 kilograms (kg) = 10 pounds (lb) | 1 liter (L) = 1 quart (qt) = 2 pints (pt) |
| 10 kilograms (kg) = 22 pounds (lb) | 2.5 liters (L) = 5 pints (pts) |
| 12 kilograms (kg) = 26 pounds (lb) | 4 liters (L) = 1 gallon (gal) |
| 50 kilograms (kg) = 110 pounds (lb) | 20 liters (L) = 5 gallons (gal) |
| 112 kilograms (kg) = 250 pounds (lb) | 50 liters (L) = 15 gallons (gal) |
| 136 kilograms (kg) = 300 pounds (lb) | 200 liters (L) = 55 gallons (gal) |
| 170 kilograms (kg) = 375 pounds (lb) | |
| 181 kilograms (kg) = 400 pounds (lb) | |

## UNITS OF LENGTH AND AREA

### U.S. to Metric Conversions

Inch (in.) = 1 inch = 25.4 millimeters (mm)
Foot (ft) = 12 inches = 0.305 meters (m)
Yard (yd) = 36 inches or 3 feet = 0.914 meters (m)
Mile (mi) = 5,280 feet = 1.609 kilometers (km)
$In^2$ (sq. in.) = 1 square inch = 6.452 square centimeters ($cm^2$)
$Ft^2$ (sq. ft.) = 144 square inches = 0.093 square meters ($m^2$)
$Yd^2$ (sq. yd.) = 1,296 square inches = 0.836 square meters ($m^2$) or 9 square feet
Acre (A) = 43,560 square feet = 0.405 hectares (ha)
$Mile^2$ (sq. mi.) = 640 acres = 2.59 square kilometers ($km^2$)

**Metric to U.S. Conversions**

Millimeters (mm) = 0.001 meters (m) = 0.039 inches (in.)
Centimeters (cm) = 0.01 meters (m) = 0.394 inches (in.)
Decimeters (dm) = 0.1 meters = 3.937 inches (in.)
Meter (m) = 1 meter = 3.281 feet (ft.)
Kilometer (km) = 1,000 meters (m) = 0.621 miles (mi.)
Square millimeter (mm$^2$) = 0.000001 square meters (m$^2$) = 0.002 square inches (in$^2$)
Square centimeter (cm$^2$) = 0.0001 square meters (m$^2$) = 0.155 square inched (in$^2$)
Square decimeter (dm$^2$) = 0.01 square meters (m$^2$) = 15.5 square inched (in$^2$)
Square meter (m$^2$) = 1 square meter (m$^2$) = 10.764 square feet (ft$^2$)
Hectare (ha) = 10,000 square meters (m$^2$) = 2.471 acres (A)
Square kilometer (km$^2$) = 1,000,000 square meters (m$^2$) = 0.386 square miles (mi$^2$)

## UNITS OF CAPACITY

### Liquid Unit Conversions from U.S. to Metric

Fluid ounce (fl oz) = 1 fluid ounce (fl oz) = 29.573 milliliters (mL)
Pint (pt) = 16 fluid ounces (fl oz) = 0.473 liters (L)
Quart (qt) = 32 fluid ounces (fl oz) or 2 pints (pt) = 0.946 liters (L)
Gallon (gal) = 6 pints (pt) or 4 quarts (qt) = 3.785 liters (L)

### Dry Unit Conversions from U.S. to Metric

Pint (pt) = 1 pint (pt) = 0.551 square decimeter (dm$^2$)
Quart (qt) = 2 pints (pt) = 1.101 square decimeters (dm$^2$)
Peck (pk) = 8 quarts (qt) = 8.810 square decimeters (dm$^2$)
Bushel (bu) = 32 quarts (qt) = 35.238 square decimeters (dm$^2$)

### Metric to Customary

Milliliter (ml) = 0.001 liters (L) = 0.034 fluid ounces (fl oz – liquid) = 0.002 pints (pt dry)
Liter (L) = 1 liter (L) = 1.057 quarts (qt liquid) = 0.098 quarts (qt dry)
Hectoliter (hL) = 100 liters (L) = 26.418 gallons (gal liquid) = 2.838 quarts (qt dry)

## UNITS OF LIQUID MEASURE

| Unit | Fluid Drams | Fluid Ounces | Quarts | Gallons | Liters |
|------|-------------|--------------|--------|---------|--------|
| Fl dram | 1 | 0.125 | 0.003906 | 0.000976 | 0.003697 |
| Fl ounce | 8 | 1 | 0.03125 | 0.007812 | 0.028413 |
| Quart | 256 | 32 | 1 | 0.25 | 0.94635 |
| Gallon | 1024 | 128 | 4 | 1 | 3.78541 |
| Milliliter | 0.2705 | 0.03381 | 0.001057 | 0.000264 | 0.001 |
| Liter | 270.0512 | 33.8140 | 1.05669 | 0.264172 | 1 |
| Cubic inch | 4.432 | 0.5541 | 0.017316 | 0.004329 | 0.016387 |

## TEMPERATURE CONVERSIONS

Fahrenheit = 9/5 °C +32
Celsius = 5/9 (°F – 32)
Kelvin = °C + 273

## Liquid Measure

1 teaspoonful = 1/6 ounce
1 tablespoonful = ½ ounce
2 pints = 1 quart
1 barrel = 31-1/2 gallons (oil, 42 gallons)
1 gallon = 8.345 pounds of water
1 cubic foot = 7.4805 gallons
1 acre foot of water = 43,560 cubic feet or 325.851 gallons

# LENGTH OF ROW PER ACRE AT VARIOUS ROW SPACINGS

| Distance between Rows (inches) | Row Length (feet/acre) |
|---|---|
| 6 | 87,120 |
| 12 | 43,560 |
| 15 | 34,848 |
| 18 | 29,040 |
| 20 | 26,136 |
| 21 | 24,891 |
| 24 | 21,780 |
| 30 | 17,424 |
| 36 | 14,520 |
| 40 | 13,068 |
| 41 | 12,445 |
| 48 | 10,890 |
| 60 | 8,712 |
| 72 | 7,260 |
| 84 | 6,223 |
| 96 | 5,445 |
| 108 | 4,840 |
| 120 | 4,356 |

# NUMBER OF PLANTS PER ACRE AT VARIOUS SPACINGS

In order to obtain other spacings, divide 43,560 (the number of square feet per acre) by the product of the between-rows and in-the-row spacings, each expressed as feet; that is, 43,560 divided by 0.75 (36 × 3 inch or 3 × 0.25 feet = 58,080).

| Spacing (inches) | Plants | Spacing (inches) | Plants | Spacing (feet) | Plants |
|---|---|---|---|---|---|
| 12 × 1 | 222,720 | 30 × 3 | 69,696 | 8 × 1 | 7,260 |
| 12 × 3 | 174,240 | 30 × 6 | 34,848 | 6 × 2 | 3,630 |
| 12 × 6 | 87,120 | 30 × 12 | 17,424 | 6 × 3 | 2,420 |
| 12 × 12 | 43,560 | 30 × 15 | 13,939 | 6 × 4 | 1,815 |

|  |  |  |  |  |  |
|---|---|---|---|---|---|
|  | 418,176 | 30 × 18 | 11,618 | 6 × 5 | 1,452 |
| 15[1] × 1 | 139,392 | 30 × 24 | 8,712 | 6 × 6 | 1,210 |
| 15 × 3 | 69,696 |  | 58,080 |  | 6,223 |
| 15 × 6 | 34,848 | 36 × 3 | 29,040 | 7 ×1 | 3,111 |
| 15 × 12 |  | 36 × 6 | 14,520 | 7 × 2 | 2,074 |
|  |  | 36 × 12 |  | 7 × 3 |  |
| 18[a] × 3 | 116,160 | 36 × 18 | 9,680 | 7 × 4 | 1,556 |
| 18 × 6 | 58,080 | 36 × 24 | 7,260 | 7 × 5 | 1,244 |
| 18 × 12 | 29,040 | 36 × 36 | 4,840 | 7 × 6 | 1,037 |
| 18 × 14 | 24,891 |  |  | 7 × 7 | 889 |
| 18 × 18 | 19,360 | 40 × 6 | 26,136 |  |  |
|  |  | 40 × 12 | 13,068 | 8 × 1 | 5,445 |
| 20[a] × 3 | 104,544 | 40 × 18 | 8,297 | 8 × 2 | 2.722 |
| 20 × 6 | 52,272 | 40 × 24 | 6,534 | 8 × 3 | 1,815 |
| 20 × 12 | 26,136 |  | 8 × 4 | 8 × 4 | 1,361 |
| 20 × 14 | 22,401 | 42 × 6 | 24,891 | 8 × 5 | 1,089 |
| 20 × 18 | 17,424 | 42 × 12 | 12,445 | 8 × 6 | 907 |
|  |  | 42 × 18 | 8,297 | 8 × 8 | 680 |
| 21[1] × 3 | 99,564 | 42 × 24 | 6,223 |  |  |
| 21 × 6 | 49,782 | 42 × 36 | 4,148 | 10 × 2 | 2,178 |
| 21 × 12 | 24,891 |  |  | 10 × 4 | 1,089 |
| 21 × 14 | 21,336 | 48 × 6 | 21,780 | 10 × 6 | 726 |
| 21 × 18 | 16,594 | 48 × 12 | 10,890 | 10 × 8 | 544 |
|  |  | 48 × 18 | 7,260 | 10 × 10 | 435 |
| 24 × 3 | 87,120 | 48 × 24 | 5,445 |  |  |
| 24 × 6 | 43,560 | 46 × 36 | 3,630 |  |  |
| 24 × 12 | 21,780 | 48 × 48 | 2,722 |  |  |
| 24 × 18 | 14,521 |  |  |  |  |
| 24 × 24 | 10,890 | 60 × 12 | 8,712 |  |  |
|  |  |  |  | 60 × 16 | 5,808 |
|  |  |  |  | 60 × 24 | 4,356 |
|  |  |  |  | 60 × 36 | 2,904 |
|  |  |  |  | 60 × 48 | 2,178 |
|  |  |  |  | 60 × 60 | 1,742 |

[a] Equivalent to double rows on beds at 30-, 36-, 40-, and 42-inch centers, respectively.

## LENGTH OF ROW PER ACRE AT VARIOUS ROW SPACINGS

| Distance between Rows (inches) | Row Length (feet/acre) |
|---|---|
| 7 | 87,120 |
| 13 | 43,560 |
| 16 | 34,848 |

| | |
|---|---|
| 19 | 29,040 |
| 22 | 26,136 |
| 23 | 24,891 |
| 25 | 21,780 |
| 31 | 17,424 |
| 37 | 14,520 |
| 42 | 13,068 |
| 43 | 12,445 |
| 49 | 10,890 |
| 61 | 8,712 |
| 73 | 7,260 |
| 85 | 6,223 |
| 97 | 5,445 |
| 109 | 4,840 |
| 121 | 4,356 |

## NUMBER OF PLANTS PER ACRE AT VARIOUS SPACINGS

In order to obtain other spacings, divide 43,560 (the number of square feet per acre) by the product of the between-rows and in-the-row spacings, each expressed as feet; that is, 43,560 divided by 0.75 (36 × 3 inch or 3 × 0.25 feet = 58,080).

| Spacing (inches) | Plants | Spacing (inches) | Plants | Spacing (feet) | Plants |
|---|---|---|---|---|---|
| 12 × 1 | 222,720 | 30 × 3 | 69,696 | 8 × 1 | 7,260 |
| 12 × 3 | 174,240 | 30 × 6 | 34,848 | 6 × 2 | 3,630 |
| 12 × 6 | 87,120 | 30 × 12 | 17,424 | 6 × 3 | 2,420 |
| 12 × 12 | 43,560 | 30 × 15 | 13,939 | 6 × 4 | 1,815 |
| | 418,176 | 30 × 18 | 11,618 | 6 × 5 | 1,452 |
| 15ᵃ × 1 | 139,392 | 30 × 24 | 8,712 | 6 × 6 | 1,210 |
| 15× 3 | 69,696 | | 58,080 | | 6,223 |
| 15 × 6 | 34,848 | 36 × 3 | 29,040 | 7 × 1 | 3,111 |
| 15 × 12 | | 36 × 6 | 14,520 | 7 × 2 | 2,074 |
| | | 36 × 12 | | 7 × 3 | |
| 18¹ × 3 | 116,160 | 36 × 18 | 9,680 | 7 × 4 | 1,556 |
| 18 × 6 | 58,080 | 36 × 24 | 7,260 | 7 × 5 | 1,244 |
| 18 × 12 | 29,040 | 36 × 36 | 4,840 | 7 × 6 | 1,037 |
| 18 × 14 | 24,891 | | | 7 × 7 | 889 |
| 18 × 18 | 19,360 | 40 × 6 | 26,136 | | |
| | | 40 × 12 | 13,068 | 8 × 1 | 5,445 |
| 20¹ × 3 | 104,544 | 40 × 18 | 8,297 | 8 × 2 | 2.722 |
| 20 × 6 | 52,272 | 40 × 24 | 6,534 | 8 × 3 | 1,815 |
| 20 × 12 | 26,136 | | | 8 × 4 | 1,361 |
| 20 × 14 | 22,401 | 42 × 6 | 24,891 | 8 × 5 | 1,089 |
| 20 × 18 | 17,424 | 42 × 12 | 12,445 | 8 × 6 | 907 |
| | | 42 × 18 | 8,297 | 8 × 8 | 680 |

| | | | | | |
|---|---|---|---|---|---|
| $21^1 \times 3$ | 99,564 | $42 \times 24$ | 6,223 | | |
| $21 \times 6$ | 49,782 | $42 \times 36$ | 4,148 | $10 \times 2$ | 2,178 |
| $21 \times 12$ | 24,891 | | | $10 \times 4$ | 1,089 |
| $21 \times 14$ | 21,336 | $48 \times 6$ | 21,780 | $10 \times 6$ | 726 |
| $21 \times 18$ | 16,594 | $48 \times 12$ | 10,890 | $10 \times 8$ | 544 |
| | | $48 \times 18$ | 7,260 | $10 \times 10$ | 435 |
| $24 \times 3$ | 87,120 | $48 \times 24$ | 5,445 | | |
| $24 \times 6$ | 43,560 | $46 \times 36$ | 3,630 | | |
| $24 \times 12$ | 21,780 | $48 \times 48$ | 2,722 | | |
| $24 \times 18$ | 14,521 | | | | |
| $24 \times 24$ | 10,890 | $60 \times 12$ | 8,712 | | |
| | | | | $60 \times 16$ | 5,808 |
| | | | | $60 \times 24$ | 4,356 |
| | | | | $60 \times 36$ | 2,904 |
| | | | | $60 \times 48$ | 2,178 |
| | | | | $60 \times 60$ | 1,742 |

[a] Equivalent to double rows on beds at 30-, 36-, 40-, and 42-inch centers, respectively.

# Reference Books and Texts

Adams, F. (Ed.). 1984. *Soil Acidity and Liming*, Number 12 in Agronomy Series. American Society of Agronomy, Madison, WI.

Adriano, D.C. 1986. *Trace Elements in the Terrestrial Environment*. Springer-Verlag, New York, NY.

Anonymous. 1991. *Best Management Practices Begin the Diagnostic Approach*. Potash & Phosphate Institute, Norcross, GA.

Anonymous. 1995. *International Soil Fertility Manual*. Potash & Phosphate Institute, Norcross, GA.

Anonymous. 2002. *Plant Nutrient Use in North American Agriculture*, PPI/PPIC/FAR Technical Bulletin 2002-1. Potash & Phosphate Institute, Norcross, GA.

Argo, W.R. and P.R. Fisher. 2001. *Understanding pH Management for Container-Grown Crops*. Meister Publication, Columbus, OH.

Barber, S.A. (Ed.). 1984. *Roots, Nutrient and Water Influx, and Plant Growth*, ASA Special Publication Number 49. American Society of Agronomy, Madison, WI.

Barber, S.A. 1995. *Soil Nutrient Availability: A Mechanistic Approach*, 2nd ed. John Wiley & Sons, Chichester, NY.

Barker, A.V. and D.J. Pilbeam. 2007. *Handbook of Plant Nutrition*. CRC Press, Taylor and Francis, Boca Raton, FL.

Belanger, R.R., P.A. Brown, D.L. Ehret, and J.G. Menzies. 1995. Soluble silicon: Its role in crop and disease management of greenhouse crops. *Plant Disease* 79:329-335.

Black, C.A. 1993. *Soil Fertility Evaluation and Control*. Lewis Publishers, Boca Raton, FL.

Bould, C., E.J. Hewitt, and P. Needham. 1983. *Diagnosis of Mineral Disorders in Plants*, Volume 1, Principles. Chemical Publishing, New York, NY.

Brown, J.R. (Ed.) 1987. *Soil Testing: Sampling, Correlation, Calibration, and Interpretation*, SSSA Special Publication 21. Soil Science Society of America, Madison, WI.

Carson, E.W. (Ed.) 1974. *The Root and Its Environment*. University Press of Virginia, Charlottesville, VA.

Epstein, E. 1994. The anomaly of silica in plant biology. *Proc. Natl. Acad. Sci.* (USA) 91:11-17.

Epstein, E. and A.J. Bloom. 2005. *Mineral Nutrition of Plants: Principles and Perspectives*, 2nd. ed. Sinauer, Sunderland, MA.

Fageria, N.K. 2009. *The Use of Nutrients in Crop Plants*. CRC Press, Boca Raton, FL.

Fageria, N.K., V.C. Baligar, and C.A. Jones. 2010. *Growth and Mineral Nutrition of Field Crops*. CRC Press, Boca Raton, FL.

Follett, R.F. (Ed.). 1987. *Soil Fertility and Organic Matter as Critical Components of Production Systems*, SSSA Special Publication Number 19. Soil Science Society of America, Madison, WI.

Glass, D.M. 1989. *Plant Nutrition: An introduction to Current Concepts*. Jones and Barlett, Publishers, Boston, MA.

Grundon, N.J. 1987. *Hungry Crops: A Guide to Nutrient Element Deficiencies in Field Crops*. Department of Primary Industries, A Queensland Government Publication, Brisbane, Australia.

Hansen, V.E., O.W. Israelsen, and G.E. Stringham. 1979. *Irrigation Principles and Practices*, 4th ed. John Wiley & Sons, New York, NY.

Hanson, A.A. 1990. *Practical Handbook of Agricultural Science*. CRC Press, Boca Raton, FL.

Havlin, J.L., S.L. Tisdale, J.D. Beaton, and W.L. Nelson. 2005. *Soil Fertility and Fertilizers: An Introduction to Nutrient Management*, 7th ed. Prentice Hall, Upper Saddle River, NJ.

Hewitt, E.J. 1966. *Sand and Water Culture Methods in Study of Plant Nutrition*. Technical Communications No. 22 (revised). Commonwealth Agricultural Bureau Horticulture and Plantation Crops. East Milling, Maidstone, Kent, England.

Hood, T.M. and J.B. Jones, Jr. (Eds.). 1997. *Soil and Plant Analysis in Sustainable Agriculture and Environment*. Marcel Dekker, Inc., New York, NY.

Jones, Jr., J. Benton. 1998. *Plant Nutrition Manual*, CRC Press, Boca Raton, FL.

Jones, Jr., J. Benton. 2000. *Hydroponics: A Practical Guide for the Soilless Grower*. St. Lucie Press, Boca Raton, FL.

Jones, Jr., J. Benton. 2001. *A Laboratory Guide for Conducting Soil Tests and Plant Analyses*, CRC Press, Boca Raton, FL.

Jones, Jr., J. Benton. 2003. *Agronomic Handbook: Management of Crops, Soils, and Their Fertility*, CRC Press, Boca Raton, FL.

Jones, Jr., J. Benton. 2011. *Hydroponic Handbook: How hydroponic growing systems work*. GroSystems, Inc., Anderson, SC.

Kabata-Pendias, A. 2010. *Trace Elements in Soils and Plants*, 4th ed., CRC Press, Boca Raton, FL.

Kalra, Y.P. (Ed.). 1998. *Handbook of Reference Methods for Plant Analysis*. CRC Press, Boca Raton, FL.

Ludwick, A.E. (Ed.). 1998. *Western Fertilizer Handbook*, 2nd ed. Interstate Publishers, Inc., Danville, IL.

Ludwick, A. E. (ed.). 1990. *Western Fertilizer Handbook* (Horticulture Edition). Interstate Publishers, Inc., Danville, IL.

Marshner, H. 1995. *Mineral Nutrition of Higher Plants*, 2nd ed. Academic Press, London, England.

Maynard, D.N. and G.J. Hochmuth. 1997. *Knott's Handbook for Vegetable Growers*, 4th ed. John Wiley & Sons, New York, NY.

Mehlich, A. 1953. *Determination of P, Ca, Mg, K, Na and NH$_4$* (mineo). North Carolina Soil Testing Division, Raleigh, NC.

Mengel, K. and E.A. Kirkby. 1987. *Principles of Plant Nutrition*, 4th. ed. International Potash Institute, Worblaufen-Bern, Switzerland.

Mills, H.A. and J.B. Jones, Jr. 1996. *Plant Analysis Handbook II*. Micro-Macro Publishing, Inc., Athens, GA.

Mortevedt, J.J. (Ed.). 1991. *Micronutrients in Agriculture*, 2nd ed., Number 4 in the Soil Science Society of America Book Series. Soil Science Society of America, Madison, WI.

Pais, I. and J.B. Jones. 1997. *Handbook of Trace Elements*. CRC Press, Boca Raton, FL.

Pessarakli, M. (Ed.). 2002. *Handbook of Plant and Crop Physiology*, 2nd ed. Marcel Dekker, Inc., New York, NY.

Pessarakli, M. (Ed.). 2010. *Handbook of Plant and Crop Stress*. CRC Press, Boca Raton, FL.

Peverill, K.I., L.A. Sparrow, and D.J. Reuter (Eds.). 1999. *Soil Analysis: An Interpretation Manual*. CSIRO Publishing, Collingwood, Australia.

Rechcigl, J.E. (Ed.). 1995. *Soil Amendments and Environmental Quality*. Lewis Publishers, Boca Raton, FL.

Reid, K. (Ed.). 1998. *Soil Fertility Handbook*, Publication 611. Ontario Ministry of Agriculture, Food and Rural Affairs, Toronto, Canada.

Rengel, Z. (Ed.). 1998. *Nutrient Use in Crop Production*. Food Products Press, New York, NY.

Resh, H.M. 2001. *Hydroponic Food Production*, 6th ed. New Concepts Press, Mahwah, NJ.

Reuter, D.J. and J.B. Robinson (Eds.). 1997. *Plant Analysis: An Interpretation Manual*. CSIRO Publishing, Collingwood, Australia.

Robson, A.D. (Ed.). 1989. *Soil Acidity and Plant Growth*. Academic Press, New York, NY.

Rubatzky, V. and M. Yamagucht. 1997. *World Vegetables: Principles, Production and Nutritive Values*, 2nd ed. International Thomson Publishing, New York, NY.

Scaife, A. and M. Turner. 1983. *Diagnosis of Mineral Disorders in Plants*, Volume 2, Vegetables. Chemical Publishing, New York, NY.

Sparks, D.L. (Ed.). 1996. *Methods of Soil Analysis, Part 3, Chemical Methods*, Number 5 in the Soil Science Society of America Book Series. Soil Science Society of America, Madison, WI.

Sprague, H. (Ed.). 1949. *Hunger Signs in Crops,* 3rd ed. David McKay Company, New York, NY.

Sumner, M.E. (Ed.). 2000. *Handbook of Soil Science*, CRC Press, Boca Raton, FL.

Swaine, D.J. 1969. *The Trace Element Content of Soils*, Commonwealth Bureau of Soil Science Technical Communications No. 48. Farnham Royal, Bucks, England.

Takahashi, E. J.F. Ma, and Y. Miyke. 1990. The possibility of silicon as an essential element for higher plants, pp. 99-122, In: *Comments on Agriculture and Food Chemistry*. Gordon and Breach Science Publications, London, Great Britain.

Tan, K.H. 1998. *Principles of Soil Chemistry,* 3rd ed. Marcel Dekker, Inc., New York, NY.

Walsh, L.M. and J.D. Beaton (Eds.). 1973. *Soil Testing and Plant Analysis*, Revised edition. Soil Science Society of America, Madison, WI.

Wang, J.Y. 1963. *Agricultural Meteorology*. Pacemaker Press, Milwaukee, WI.

Westerman, R.L. 1990. *Soil Testing and Plant Analysis*, 3rd ed., Number 3 in the Soil Science Society of America Book Series. Soil Science Society of America, Madison, WI.

Westervelt, J.D. and H.F. Reitz. 2000. *GIS in Site-Specific Agriculture*. Interstate Publishers, Danville, IL.

Wichmann, W. (ed.). 1992. *World Fertilizer Use Manual*. International Fertilizer Industry Association, Paris, France.

Wolf, B. 2000. *The Fertile Triangle: The Relationship of Air, Water, and Nutrients in Maximizing Productivity*. Food Products Press, Binghamton, NY.

# References

Anonymous. 1991. *Best Management Practices Begin the Diagnostic Approach*. Potash & Phosphate Institute, Norcross, GA.

Anonymous. 1992. *Handbook on Reference Methods for Soil Analysis*. Soil and Plant Analysis Council, Inc., Athens, GA.

Anonymous. 1997. Water: What's in it and how to get it out. *Today's Chemist* 6(1):16–19.

Anonymous. 1999. *Soil Analysis Handbook of Reference Methods*. CRC Press, Boca Raton, FL.

Arnon, D.I. and P.R. Stout. 1939. The essentiality of certain elements in minute quantity for plants with special reference in copper. *Plant Physiol.* 14:371–375.

Asher, C.J. 1991. Beneficial elements, functional nutrients, and possible new essential elements, pp. 703–723. In J.J. Mortvedt (Ed.), *Micronutrients in Agriculture*, SSSA Book Series Number 4, Soil Science Society of America, Madison, WI.

Belanger, R.R., P.A. Brown, D.L. Ehret, and J.G. Menzies. 1995. Soluble silicon: Its role in crop and disease management of greenhouse crops. *Plant Disease* 79:329-335.

Beverly, R. 1991. A Practical Guide as the Diagnostic and Recommendation Integrated System (DRIS). Micro-Marco Publishing, Athens, GA.

Bray, R.H. and L.T. Kurtz. 1945. Determination of total, organic and available forms of phosphorus in soils. *Soil Sci.* 59:35–45.

Brown, P.H., R.M. Welsh, and E.E. Cary. 1987. Nickel: A micronutrient essential for higher plants. *Plant Physiol.* 85:801–803.

Carson, E.W. (Ed.). 1993. I. University Press of Virginia, Charlottesville, VA.

Carson, R. 1962. *Silent Spring*. Houghton Mifflin Company, Boston, MA.

Carter, R.L. (Ed.), 1993. *Soil Sampling and Method of Analysis*. CRC Press, Boca Raton, FL.

Cooper, A. 1976. *Nutrient Film Technique for Growing Crops*. Grower Books, London, England.

Epstein, E. 1965. Mineral Nutrition, pp. 438–466. In J. Bonner and J.E. Varner (Eds.), *Plant Biochemistry*. Academic Press, Orlando, FL.

Epstein, E. 1994. The anomaly of silica in plant biology. *Proc. Natl. Acad. Sci.* (USA) 91:11-17.

Eastwood, T. 1947. *Soilless Growth of Plants*, 2nd ed. Reinhold Publishing Company, New York.

Fitts, J.W. and L.B. Nelson. 1956. The determination of lime and fertilizer requirements of soils through chemical tests. *Adv. Agron.* 8:241–282.

Folds, Evan. 2009. The dynamic nature of water. Part II. *Maximum Yield* (April 2009):70–72.

Grunden, N.J. 1987. *Hungry Crops: A Guide to Nutrient Element Deficiencies in Field Crops*. Department of Primary Industries, Queensland Government Publication, Brisbane, Australia.

Hayden, A.L. 2003. The good, the bad, and the ugly: Inorganic fertilizers, toxic metals and proposed regulations. *The Growing Edge* 15(2):42–50.

Hewitt, E.J. 1966. *Sand and Water Culture: Methods Used in Study of Plant Nutrition*. Technical Communication No. 22 (revised). Commonwealth Agricultural Bureau of Horticulture and Plantation Crops, East Milling, Maidstone, Kent, England.

Hoagland, D.R. and D.I. Arnon. 1950. *The Water Culture Method for Growing Plants without Soil*. Circular 347. California Agricultural Experiment Station, University of California, Berkeley, CA.

Houba, V.J.G., E.J.M. Temmingghoff, G.A. Gaikhorst, and W. van Eck. 2000. Soil analysis procedures using 0.001M calcium chloride extraction reagent. *Commun. Soil. Sci. Plant Anal.* 31:1299–1396.

Jones, Jr., J. Benton. 1998. *Plant Nutrition Manual*, CRC Press, Boca Raton, FL.

Jones, Jr., J. Benton. 2001. *Laboratory Guide for Conducting Soil Tests and Plant Analysis*. CRC Press, Boca Raton, FL.

Jones, Jr., J. Benton. 2005. *Hydroponics: A Practical Guide for the Soilless Grower*. 2nd ed. CRC Press, Boca Raton, FL.

Jones, Jr., J. Benton. 2011. *Hydroponic Handbook: How hydroponic growing systems work*. GroSystems, Inc., Anderson, SC.

Kabata-Pendias, A. and H. Pendias. 2000. *Trace Elements in Soils and Plants*, 3rd ed. CRC Press, Boca Raton, FL.

Ludwick, A. E. (Ed.). 1990. *Western Fertilizer Handbook* (Horticulture Edition). Interstate Publishers, Inc., Danville, IL.

Markert, B. 1994. Trace element content of "Reference Plant." In D.C. Adriano, Z.S. Chen, and S.S. Yang (Eds.), *Biochemistry of Trace Elements*. Science and Technology Letters, Northwood, NY.

Maynard, D.N. and G.J. Hochmuth 1997. *Knott's Handbook for Vegetable Growers*, 4th ed. John Wiley & Sons, New York.

Mehlich, A. 1953. *Determination of P, Ca, Mg, K, Na and NH₄* (mineo). North Carolina Soil Testing Division, Raleigh, NC.

Mehlich, A. 1984. Mehlick No. 3 soil test extractant: A modification of the Mehlich No. 2. extractant. *Commun. Soil Sci. Plant Anal*. 15:1409–1416.

Mertz, W. 1981. The essential trace elements. *Science* 213:1332–1338.

Mills, H.A. and J.B. Jones, Jr. 1996. *Plant Analysis Handbook II*, Micro-Macro, Athens, GA.

Morgan, L. 2000. Beneficial elements for hydroponics: A new look at plant nutrition. *The Growing Edge* 11(3):40–51.

Morgan, L. 2003. Hydroponic substrates. *The Growing Edge* 15(2):54–66.

Morgan, L. 2005. It's the water. *The Growing Edge* 16(5):46–48.

Morgan, M.F. 1941. *Chemical Diagnosis by the universal Soil Testing System*. Conn. Agr. Exp. Sta. Bul. 459. New Haven, CT.

Nelson, W.L., J.W. Fitts, L.D. Kardos, W.T. McGeorge, R.Q. Parks, and J.F. Reed. 1951. *Soil Testing in United States*, 0-979953. National Soil and Fertilizer Research Committee. U.S. Government Printing office, Washington, DC.

Neilson, F.H. 1984. Nickel. In *Biochemistry of the Essential Ultratrace Elements*, E. Frieden (Ed.), Plenum Press, New York, pp. 293–308.

Olsen, S.R., C.V. Cole, F. Watanabe, and L.A. Dean. 1954. *Estimation of the Available Phosphorus in Soils by Extraction with Sodium Bicarbonate*. USDA Circular No. 939. U.S. Government Printing Office, Washington, DC.

Pais, I. 1992. Criteria of essentiality, beneficiality, and toxicity of chemical elements. *Acta Aliment*. 21(2):145–152.

Pallas, Jr., J.E. and J.B. Jones, Jr. 1978. Platinum uptake by horticultural crops. *Plant and Soil* 50:207–212.

Parnes, R., 1990. *Fertile Soil: A Grower's Guide to Organic & Inorganic Fertilizers*. AgAccess, Davis, CA.

Porter, J.R. and D.W. Lawlor (Eds.). 1991. *Plant Growth Interaction with Nutrition and Environment*. Society for Experimental Biology. Seminar 43. Cambridge University Press, New York.

Resh, H.M. 2001. *Hydroponic Food Production*, 6th ed. New Concepts Press, Mahwah, NJ.

Reuter, D.J. and J.B. Robinson, Jr. (Eds.). 1997. *Plant Analysis: An Interpretation Manual*, CSIRO Publishing, Collingwood, Australia.

Schollenberger, C.J. and R.H. Simon. 1945. Determination of exchangeable carapcity and exchangeable bases in ammonium acetate method. *Soil Sci*. 59:13–24.

Sheldrake, R., Jr. and J.W. Boodley. 1965. *Commercial Production of Vegetable and Flower Plants*. Cornell Extension Bulletin 1065. Cornell University, Ithaca, NY.

Soltanpour, P.N. and A.P. Wokman. 1979. A new soil test for simultaneous extraction of macro- and micro-nutrients in alkaline soils. *Comm. Soil Sci. Plant Anal*. 8:195–207.

Sumner, M.E. (Ed.). 1980. *Handbook of Soil Science*. CRC Press, Boca Raton, FL.

Takahashi, E. and Y. Miyake. 1977. Silica and Plant Growth, pp. 603–604. In *Proceedings International Seminar Soil Environmental and Fertility Management Intensive Agriculture*. National Institute of Agricultural Science, Tokyo, Japan.

Takahashi, E. J.F. Ma, and Y. Miyake. 1990. The possibility of silicon as an essential element for higher plants, pp. 99-122, In: *Comments on Agriculture and Food Chemistry*. Gordon and Breach Science Publications, London, England.

Wichmann, W. (Ed.). 1992. *World Fertilizer Use Manual*. International Fertilizer Industry Association, Paris, France.

Wolf, B. 1982. An improved universal extracting solution and its use for diagnosing soil fertility. *Commun. Soil Sci. Plant Anal.* 13:1005–1013.

# Index

Printed in the United States
by Baker & Taylor Publisher Services